Recent Technologies in Capture of CO$_2$

Edited By

Rosa-Hilda Chavez

National Institute of Nuclear Research
(Instituto Nacional de Investigaciones Nucleares, Mexico)
Environmental Sciences Management
Environmental Studies Department
Mexico

Co-Editor

Javier de J. Guadarrama

Toluca Institute of Technology
(Instituto Tecnologico de Toluca, Mexico)
Electrical and Electronics Engineering Department
Mexico

Bentham Science Publishers
Executive Suite Y - 2
PO Box 7917, Saif Zone
Sharjah, U.A.E.
subscriptions@benthamscience.org

Bentham Science Publishers
P.O. Box 446
Oak Park, IL 60301-0446
USA
subscriptions@benthamscience.org

Bentham Science Publishers
P.O. Box 294
1400 AG Bussum
THE NETHERLANDS
subscriptions@benthamscience.org

CONTENTS

FOREWORD

Global warming is a result of increasing anthropogenic CO_2 emissions, and the consequences will be dramatic climate changes if no action is taken. One of the main global challenges in the years to come is to reduce the CO_2 and GHG emissions.

It is widely accepted that the climate change issue is a global concern that directly links with the emission of greenhouse gases – notably carbon dioxide (CO_2) associated with the use of fossil fuels around the world. Last decade 32.6 Gt of CO_2 was released to the atmosphere as a consequence of the global harnessing of fossil fuels. From this amount 41% of the CO_2 relates to coal. From a more simplistic point of view the challenge may be construed as just obviating these emissions – either by swapping to carbon-neutral primary energy – possibly wind, biomass, photovoltaic and nuclear, or by isolating the CO_2 and preventing it from being released to the atmosphere (or combined). But the challenge, also involves a geopolitical dimension represented by a fast-growing global population and increasing prosperity level. This dimension is prone to justify the extended electrification in developing nations. On this basis prognoses suggest that the global electricity demand will grow by a factor 3 to 7 within this century.

So far, no sustainable primary energy has been identified as being capable of supplying electric power – on demand - in very large quantities at a reasonable cost the global primary energy supply will continually be dominated by fossil fuels over the foreseeable future - regardless of the CO_2 emissions that constitute a major drawback of these fuels.

The previous considerations are behind developing a new book on this topic and caused the motivation. The book edited by Prof. Chavez provides a broad overview on several techniques feasible to developed and reduce the amount of CO_2 released into the atmosphere.

We are most grateful to the numerous authors who put so many times into writing outstanding chapters. This book includes all the relevant and current information on:

- Expand knowledge and research to establish criteria and solutions to reduce of

- GHG emissions.

- Awakening interest of new generations in knowledge and research in this area.

- Establish links between researchers, as well as in the academic community.

- Establish solutions to this huge climate change in the world.

Much information is completely updated. Although we believe that this book will provide a clear understanding of these techniques, we invite our readers to inform us about differences of opinion they may have with its contents and areas that need improvement.

Lydia Paredes Gutierrez
General Manager
Instituto Nacional de Investigaciones Nucleares
Mexico

PREFACE

As the Editor of *Recent technologies in capture of CO₂,* it is a great pleasure to introduce this E-Book to readers, researchers, future authors, and colleagues. The decision by Bentham to launch this E-Book is very timely. It is an honor that so many distinguished scientists have been willing to support this initiative. The aim of *Recent technologies in capture of CO₂* is to provide a unique, multidisciplinary platform for the dissemination of leading edge research work on CO_2 capture and minimizing of environmental impact of carbon dioxide. This E-Book covers the vibrant and diverse field of carbon dioxide capture, from the basic science and the development of new processes, technologies and applications.

Part I. Oxicombustion

Chapter 1 by Prof. Luis M. Romeo and co-authors on **Oxy-fuel combustion in fluidized beds**, presents Fluidized beds with particular features driving this technology as an appropriate candidate technology to apply oxy-fuel combustion, producing a highly concentrated CO_2 flue gas stream to be processed and stored. Still, there are several issues differentiating the conventional combustion to that with O_2/CO_2 mixtures. This chapter examines the main issues involved in oxy-fuel combustion in fluidized beds through the experimental results obtained in the CIRCE oxy-fuel bubbling fluidized bed.

Part II. Post-Combustion

Chapter 2 by Prof. Li Zhao and Prof. Ludger Blum on **Gas Separation membranes used in post-combustion capture** describes mass and energy balances for single-stage and multi-stage membrane systems used in coal-fired power plant. In order to evaluate different membrane capture concepts, a comparison with chemical absorption process was carried out, considering different degrees of CO_2 separation. Furthermore, a cost model was developed to make a further analysis of the optimized concept in view of the tradeoff balance between material and energy consumption. The correlation between the membrane parameters (selectivity, permeability) and capture cost was investigated.

Chapter 3 by Prof. Martin Picon and co-authors on **Minimizing energy consumption in Co_2 capture processes through process integration** focuses on the thermal integration for minimum external energy consumption of CO_2 capture process using amines. Post combustion capture of CO_2 through the use of amines is a well established technique; the stand alone process is highly energy intensive since the recovery of the amine solution is achieved through the use of a separation process where heating and cooling are required. Energy integration of the hot flue gas coming from a power station plant from where CO_2 is absorbed can serve the purpose of providing the heating and cooling needs to the process. Heat recovery through steam rising is considered for heating, cooling and for the production of power for the operation of pumps and compressors. The results show that the needs of the largest energy user of the process can be fully met by heat integration.

Chapter 4 by Prof. Rosa-Hilda Chavez and co-authors on **Characterization and application of structured packing for CO_2 capture** evaluates the minimum energy consumption for solvent regeneration and maximum CO_2 absorption with 600 t/hr flue gas flow simulated by *Aspen Plus*™ of CO_2 capture process, using Monoethanolamine (MEA) at 30 weight%. The parameters studied were: 1) energy consumption at reboiler of stripper, 2) absorption separation efficiency, 3) flow ratio (L/G) in order to find the load or turbulence regimen in absorption process, and 4) absorption and stripper column diameters at different flue gas flows. This work contributes structured packing study in separation columns, like: ININ 18, *Sulzer* BX and *Mellapak* 250Y, and the advance of CO_2 capture technology. Hydrodynamic and mass transfer models were used to evaluate pressure drops and height of mass transfer equivalent unit, per each packing. The results showed that *Sulzer* BX has the highest volumetric mass transfer coefficient values and the lowest height of mass transfer equivalent unit, with $3.76s^{-1}$ and $0.316m$, respectively, and the most absorption efficiency with 89.17% in comparison with respect to the other two packings with 600 ton/hr flue gas flow treated.

Chapter 5 by Prof. Luis M. Romeo and co-authors on **Looping cycles for CO_2 capture** introduces to the reader the calcium looping process for CO_2 capture. This technology makes use of the idea that lime may be reused in a cyclic process to remove CO_2 from a mixture of gases where carbonate is calcined to generate a pure stream of CO_2 ready for sequestration. This chapter analyses the energy penalties of the Ca-looping CO_2 capture system, different types of sorbents and their performance subjected to repeated cycles of carbonation and calcination, the CO_2 capture efficiency and the possibility of integration of Ca-looping and power plants to reduce energetic penalties.

Chapter 6 by Prof. Rosa-Hilda Chavez and co-authors on **Liquid-gas contactors material properties for CO_2 capture in absorption column** discusses some structural characteristics and properties of three regular packing materials: metallic, polymeric and ceramic; in order to select the best one to capture CO_2 in an absorption column. The study was conducted by making the following tests: geometric physical properties such as wetted area and porosity; mechanical properties like, stress, hardness, modulus and compression resistance, structure and microstructure morphologic, chemical composition, rate of corrosion in electrochemical cell in medium of 1N of H_2SO_4 and Monoethanolamine (MEA) at 30% in aqueous solution by using standard procedures of the American Society of Testing Materials (ASTM) and own developed procedures for the equipment used. The structures of materials also were evaluated by X-ray diffraction and the surface of the material by scanning electron microscopy. It was concluded that metallic material is suitable in CO_2 gas treatment because it presented lower etching of 1N of H_2SO_4, and MEA at 30% in aqueous solution and giving the most absorption of CO_2.

Part III. Chemical Kinetics

Chapter 7 by Prof. Felipe Bustamante and co-authors on **Rate-based models and design of packed columns for absorption of carbon dioxide** shows details on the selection of the equipment for CO_2 absorption, compares rate-based models (RBM) and equilibrium models (EQM) for absorption of CO_2 in ammonia, and describes the development of a RBM for reactive absorption. From the literature reviewed and conditions prevailing in the absorption process, it is advantageous to carry out the process in packed columns. The comparison of EQM and RBM showed differences in temperature and concentration profiles along a 10 stages packed column. The temperature profile in EQM shows that the liquid temperature matches the inlet gas temperature at the fourth stage and remains constant until the bottom of the column. From the RBM, on the other hand, three temperature profiles were obtained: interface, gas, and liquid streams; temperatures in the EQM were almost 10°C higher than in the RBM. For intermediate stages, the EQM predicted higher mol fraction of CO_2 than RBM.

Part IV. Miscellanies

Chapter 8, by Prof. Juan Aspiazu and Arturo Aspiazu on **Determination of Carbon and Oxigen captured from environmental samples** describes some efficient physicochemical procedure for either CO_2 dissociation or recycling. Ion accelerators provide a suite of techniques, collectively referred to as IBA, offering excellent options for the analysis of this kind of contamination.

Finally, I sincerely hope that this eBook on *Recent technologies in capture of CO₂* will be adopted by readers and authors, and that it will be able to make an important contribution to the development of CO_2 capture development community and the scientific achievements in our field.

Rosa-Hilda Chavez

National Institute of Nuclear Research
(Instituto Nacional de Investigaciones Nucleares, Mexico)
Environmental Sciences Management
Environmental Studies Department
Mexico

&

Javier de J. Guadarrama

Toluca Institute of Technology
(Instituto Tecnologico de Toluca, Mexico)
Electrical and Electronics Engineering Department
Mexico

List of Contributors

Aspiazu, J.A.
National Institute of Nuclear Research, Accelerators Department, La Marquesa, Ocoyoacac, 52750, Mexico

Aspiazu, A.
National Autonomous University of Mexico, Faculty of Chemistry, Av. Insurgentes Sur 4411 Ed 25-301, Tlalpan D.F., 14430, Mexico

Bolea, I.
CIRCE Institute - University of Zaragoza, 50018-Zaragoza, Spain

Bustamante, F.
Environmental Catalysis Research Group, Chemical Engineering Department, Universidad de Antioquia, Colombia

Chavez, R.H.
National Institute of Nuclear Research, Carretera Mexico-Toluca S/N, La Marquesa, Ocoyoacac, 52750, Mexico

Guadarrama, J.
Toluca Institute of Technology, Metepec, 52140, Mexico, Mexico

Guedea, I.
CIRCE Institute - University of Zaragoza, 50018-Zaragoza, Spain

Hoyos, A.E.
Chemical Engineering Department, Universidad de Antioquia, Colombia

Lara, Y.
CIRCE Institute - University of Zaragoza, C/Marino Esquillor Gomez, 15, 50018-Zaragoza, Spain

Lisbona, P.
CIRCE Institute - University of Zaragoza, C/Marino Esquillor Gomez, 15, 50018-Zaragoza, Spain

Lupiañez, C.

CIRCE lnstitute - University of Zaragoza, C/Marino Esquillor Gomez, 15, 50018-Zaragoza, Spain

Martínez, A.

CIRCE lnstitute - University of Zaragoza, C/Marino Esquillor Gomez, 15, 50018-Zaragoza, Spain

Müller, A.

Laboratory of Fluid Separations, Department of Biochemical and Chemical Engineering, TU Dortmund University, Denmark

Picón, M.

Department of Chemical Engineering, University of Guanajuato, Guanajuato, Gto.36050, México

Romeo, L.M.

CIRCE lnstitute - University of Zaragoza, C/Marino Esquillor Gomez, 15, 50018-Zaragoza, Spain

Salazar, A.

National Institute of Nuclear Research, La Marquesa, Ocoyoacac, 52750, Mexico

Torres, A. E.

Department of Chemical Engineering, University of Guanajuato, Guanajuato, Gto.36050, México

Villa, A.L.

Environmental Catalysis Research Group, Chemical Engineering Department, Universidad de Antioquia, Colombia

Zhao, L.

Institute of Energy and Climate Research, Fuel Cells (IEK-3), Leo-Brandt-Straße, Forschungs zentrum Jülich, D-52425 Jülich, Germany

2

CHAPTER 1

Oxy-Fuel Combustion in Fluidized Beds

Isabel Guedea, Irene Bolea, Carlos Lupiañez, Luis M. Romeo[*] and Luis I. Díez

CIRCE Institute - University of Zaragoza, C/Marino Esquillor Gomez, 15, 50018-Zaragoza, Spain

Abstract: Fluidized beds particular features drive this technology as an appropriate candidate technology to apply oxy-fuel combustion, producing so a highly concentrated CO_2 flue gas stream to be processed and stored. Still, there are several issues differencing the conventional combustion to that with O_2/CO_2 mixtures. This chapter examines the main issues involved in oxy-fuel combustion in fluidized beds through the experimental results obtained in the CIRCE oxy-fuel bubbling fluidized bed.

The fluidization velocity during oxy-firing is in general, below the air-firing case. This is caused by the higher O_2 concentration in the oxidant stream, together with the higher gas density when substituting air-N_2 by CO_2. The lower fluidization velocity affects in opposite ways the combustion and pollutant formation: it increases the residence time of particles in bed, whereas poorer mixing of fuel particles disadvantages reactions. Higher O_2 at inlet obliges to increase the fuel input to maintain proper fluid-dynamics conditions. This is adequate for the combustion efficiency and also, for the in-furnace SO_2 capture. Unlike in conventional combustion, SO_2 capture optimum temperature is higher than 850 ºC, because of the influence of high CO_2 partial pressure.

NO_x emissions showed no significant differences with air-firing case, if concentration is expressed per unit of energy. This is due again to the lower flow of flue gases per thermal fuel input. Plant heat balance of large oxy-fuel fluidized bed boilers will change considerably in oxy-fuel case. Boilers will be more compact and thus, additional heat transfer surface will be essential.

Keywords: Oxyfuel, SO_2 and NO_x emissions, combustion efficiency, fluidized bed, desulfurization, heat transfer in fluidized beds, pilot plants, power plant.

1. INTRODUCTION

Oxy-fuel combustion is one of the three groups of technologies (along with the

*Corresponding author Luis M. Romeo: CIRCE Institute - University of Zaragoza, C/Mariano Esquillor Gomez, 15, 50018-Zaragoza, Spain; Tel: +34 976 76 25 70; Fax: +34 976 73 20 78; E-mail: luismi@unizar.es

so-called pre-combustion and post-combustion) aiming to capture CO_2 from large stationary emissions sources. In oxy-fuel combustion, a mixture of pure oxygen diluted with a recycled part of the flue gas (RFG), mainly composed of CO_2, substitutes the conventional air. Water content in exhaust gases can be easily condensed out and thus a high CO_2 concentration is reached for further purification, transport and storage.

Oxy-fuel combustion is not a novel concept. Metallurgy and glass industries use pure or enriched oxygen atmospheres in ovens and manufacturing processes since long ago [1-3]. Oxy-fuel combustion with the purpose of capturing CO_2 was first proposed in 1981 and early studies were carried out at Argonne National Laboratory [4, 5]. The main reason was its potential for enhanced oil recovery, adding value to the CO_2 stream. However, the oil crisis diminished this interest for CO_2 capture. In the late 1990s, attention grew again towards oxy-fuel technology, due to the imminent necessity of looking for solutions devoted to control greenhouse gas emissions.

Currently, fossil fuels account for more than 60% of the electricity generation. Coal has been a key component of the electricity generation mix worldwide. It fuels more than 30% of the world's electricity, although this share is higher in some areas: South Africa, 93%, Poland, 92%, China, 79%, India, 69% and United States, 49% [6]. The combustion of coal is mostly carried out in pulverized coal (PC) boilers. Fluidized bed technology, however, has gained an outstanding importance in the last years. In 2009, the largest CFB plant of 460 MW was commissioned in Lagisza (Poland). The main driver for starting the development programs lied on the abundant quantity of low-grade coals with high sulphur content in countries like UK, USA, China and Germany. Works carried out by EERC/ANL (3 MW), IFRF (2.5 MW), B&W (1.5 MW) and CANMET (0.3 MW), confirmed the feasibility of oxy-fuel combustion in PC boilers.

Although the ratio of CFB combustors over the PC for conventional air-firing was around one/fourth in North America and Europe in 2007 [7], the oxy-fuel pilot and demonstration roadmap shows that both technologies are considered equally suitable. In fact, the two options are already running at demonstration scales. The first 30 MW oxy-fuel PC was commissioned in 2008 in Schwarze Pumpe (Germany) and the 30 MW oxy-fuel CFB started to operate in CIUDEN (Spain) in 2011 [8]. Table **1** summarized the largest (>15 MW) research projects on oxy-combustion worldwide and those programmed at demonstration scale.

Table 1: Pilot and demonstration plants of oxy-fuel combustion [8-16]

Project/ Institution	Country	Size (MW$_{th}$)	Boiler Type	Fuel	CO$_2$ Capture Details
Babcock & Wilcox	USA	30	Pilot PC	Bit. SubBit, Lig	No Capture
Jupiter	USA	15	PC retrofit. No FGR	NG, High sulphur coal	25% flue gases to be treated. No storage
Doosan Babcock	UK	40	Pilot PC		
Vattenfall	Germany	30	Pilot PC	Lig. Bit.	With CCS
Total, Lacq	France	30	Industrial. Steam for utilities	NG and Liq. Fuels	With CCS in gas depleted reservoir
Enel	Italy	48	Presurized PC		
Jupiter (Pearl Plant)	USA	15	5 MWe PC	Bit	Side stream
Callide	Australia	~90	30 MWe Retrofit PC	Bit	75 tpd to CCS. Road transport to a depleted gas field
Ciuden-PC	Spain	20	Pilot PC	Antr. Pet cok	With CCS
Ciuden-CFB	Spain	30	Pilot CFB	Antr. Pet cok	With CCS

The two main distinguishing features of FB technology are related to its capacity of burning difficult fuels with high moisture, ash, sulphur proportions or costly pretreatment requirements. FB allows low records for SO$_2$ emissions by the addition of calcium based sorbents inside the boiler, contributing to the use of low-rank local high sulphur coals. Regarding NO$_x$ emissions, bed temperatures allow reducing these emissions considerably, compared to PC technology. Both issues approaches fluidized bed combustion to the concept of "clean coal technology". FB arises then as an outstanding opportunity for applying oxy-fuel combustion. This will be a further step towards develop a realistic concept of "zero-emissions coal power plants".

With this global picture, research must focus on overcoming the barriers to make oxy-fuel technology a reliable option for the new power plants generation. Table **2** summarizes the main research groups investigating on oxy-fuel pilots plants. One of the main concerns of scientific community is precisely coping with the pollutant emissions, during oxy-firing, inherent in solid fuels combustion, taking advantage of the FB features for capturing SO$_2$ and avoiding part of the NO$_x$ formation. Some of the groups in Table **2** had previous experiences on these topics in air-firing and retrofitted existing plants to oxy-fuel operation, like in CANMET, Utah or VTT.

Inherent particularities of fluidized beds are related to the movement of particles. They provide the outstanding characteristic of uniform and controlled temperature. Temperature is the most important operating parameter, influencing not only the pollutant formation and reduction, but also the combustion efficiency. Due to O$_2$ presence and high CO$_2$ concentration, it will also determine particle conversion mechanisms and the differences with the known in air-firing case.

At large scale, some modeling results have been published up to date, to predict future design of large oxy-fuel CFB [17-19]. More compact boilers and higher fuel input per boiler wall surface are common conclusions from these simulations.

Table 2: Research groups with experimental experiences on oxy-fuel fluidized bed pilot plants

Group	Experimental Facility (Type/ID/Height)	Modeling	Refs.
CANMET	CFB/100 mm/5 m	No	[20]
VTT	CFB/167 mm/8 m	Yes, dynamic simulation	[21]
Czestochowa University of Technology	CFB/250 mm/5 m	Yes, at large scale (not validated)	[22, 23]
ALSTOM	CFB/660m-1000mm/18 m	Yes, at large scale (not validated)	[17]
CSIC-ICB	BFB/100 mm/0.6 m	No	[24]
University of Utah	CFB/250 mm/6.4 m	No	[25]
METSO	CFB/1 x 1 m^2/13 m	Yes	[26, 27]
Wien Technical University	CFB/150 mm/5 m	No	[28]
CNR	BFB/40 mm/1 m	Yes	[29]
Southeast University	CFB/122-150 mm/ 4.2 m	No	[30]
CIRCE	BFB/207 mm/2.5 m	Yes	[31-33]

This chapter reviews the late findings on oxy-fuel combustion in fluidized beds and the main issues related to the combustion performance pollutant emissions and heat balance, paying attention to the main differences between air firing and oxy-firing. The influence of the main parameters during plant operation is also tackled.

Experimental values shown along this chapter were obtained in the CIRCE oxy-fuel bubbling fluidized bed pilot plant, see Fig. **1**. The combustor is 0.207 m internal diameter; a freeboard height was fixed at 1.8 m with an estimated maximum dense-bed height of 0.7 m. Two hoppers discharging into a mixing screw-feeder allow introducing wide range of fuel/sorbent/biomass blends. During oxy-fuel operation, oxygen and carbon dioxide mixture is fed from commercial gas canisters. A recirculation fan allows substituting commercial CO$_2$ by recirculation flue gases (RFG). The bed is water-cooled during stable operation by four independent cooling jackets at four heights, while the freeboard is refractory lined.

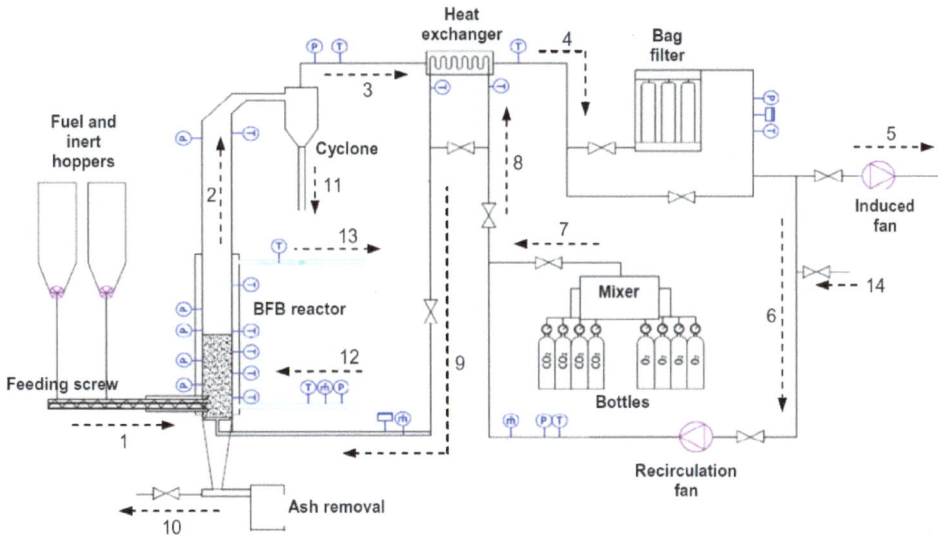

1	Solids feeding to the reactor	8	Cold comburent to heat exchanger
2	Flue gas leaving the reactor	9	Hot comburent to the reactor
3	Flue gas leaving the cyclone	10	Ash from the reactor bottom-pit
4	Cooled flue gas to bag filter	11	Ash from the cyclone leg
5	Flue gas to the stack	12	Cooling water inlet
6	Recycled flue gas	13	Cooling water outlet
7	O$_2$/CO$_2$ mixtures from the bottles	14	Air inlet (when conventional firing)

Figure 1: Diagram of the reactivity under different temperature [32].

2. FUEL CONVERSION IN OXY-FUEL

2.1. Particle Combustion

Particle conversion is expected to be modified by the different oxidizing atmosphere in comparison to conventional air-firing: both devolatilization and char combustion can be affected; the extent has to be determined.

As the fuel enters in the hot environment of the combustion chamber, it heats up and most of the moisture releases at a temperature near the water boiling point (*drying*). While the fuel temperature increases, the volatile fraction vaporizes (*devolatilization*). The volatile gas mixes with oxygen, ignites and burns in a flame created around the particle. Heat from the flame further rises the particle temperature and the rate of devolatilization increases. The remaining part of the fuel, the porous char, reacts with oxygen that previously diffused to the surface (*char combustion*). This is a slower process than drying and devolatilization and it constitutes most of the time required for a fuel particle to be fully burnt. The reaction rate depends on oxygen, temperature and char reactivity.

Fuel combustion is largely affected by fuel reactivity, *i.e.* the ability to react with oxygen and the rate at which this process occurs. Knowing the fuel reactivity is essential for the plant operation. It influences decisive issues as the combustion efficiency, the temperature profiles, or the arrangement of heat transfer surfaces.

The fuel reactivity depends both on physical and chemical properties: the size and internal structure, the temperature and the oxygen concentration in the particle surrounding. Under oxy-fuel combustion fuel reactivity will be affected by the change on gas composition, particularly by the presence of high CO_2 concentration in the combustion atmosphere.

When the fuel burns in an O_2/CO_2 atmosphere O_2 diffusivity lowers. Particle conversion is governed by surrounding O_2, temperature and particle size. The diagram in Fig. **2** schematizes the influence of temperature on the mechanism that limits the fuel combustion: diffusion, kinetic or both. Higher temperature promotes higher reactivity and thus, O_2 diffusion is the limiting mechanism. The diffusion is the slowest process in a wider range of conditions due to the lower diffusivity of all the gases in CO_2 [34]. By an exhaustive analysis of the char morphology, Pohlman [35] proved that under O_2/N_2 atmosphere kinetics controls the conversion and diffusion governs the process under O_2/CO_2.

The majority public studies up to now, reporting the behavior of the fuel particle under O_2/CO_2 blends, focus on pulverized coal (PC) boilers. Unlike in FB boilers, in PC the particle diameter is smaller and the temperature higher than in FB, which means that kinetics, governs the combustion. The diffusion gains importance for FB boiler, because of the lower temperature and larger particles.

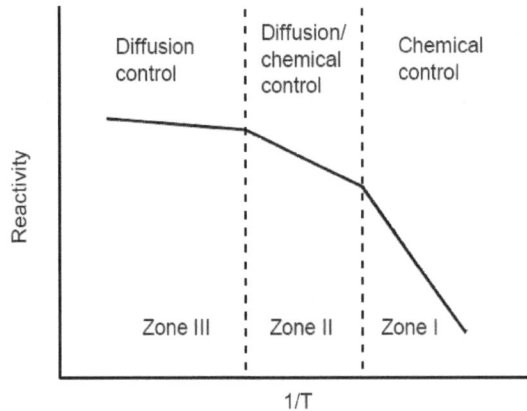

Figure 2: Diagram of the reactivity under different temperature.

In PC boilers, it is not clear the effect of O_2/CO_2 atmosphere on the devolatilization step. Some authors, like Rathman *et al.,* [36], detected a major quantity of volatile matter in CO_2 atmospheres. They found that gasification reactions promoted by higher temperatures could cause this increase. However, other studies [37, 38] did not appreciate any devolatilization differences between O_2/N_2 and O_2/CO_2 atmospheres. The only study [39] carried out in a fluidized bed in the literature highlights the higher volatile matter released under oxy-firing conditions.

As concerns char conversion, the experimental findings [36, 37] agree on pointing out the burning time and the particle temperature as the main differences between air and oxy-fuel. For O_2 concentration similar to air, burning time resulted from 44 to 80% longer. The coal ignition delays under oxy-fuel conditions, promoted by the lower O_2 diffusivity and by the lower adiabatic flame temperature when CO_2 is present [40]. Different explanations can be found for this effect. While some authors found the reason in the lower diffusivity of gaseous components in CO_2 compared to N_2, others found it in the higher thermal capacity of CO_2. However, by increasing the oxygen concentration up to 30%, the ignition temperature and the burning time turn out similar for air and O_2/CO_2 atmospheres. Higher O_2 concentrations will augment the burning rate and the particle temperature. Another important observation by several authors was that higher volatile coals showed closer values for time burnout in both atmospheres [36, 41].

Even though some of the results obtained for PC conditions could be also applied to FB, specific research has to be done due to the differences in the particle sizes and reactor conditions in a FB. The scarce literature of FB up to now [39, 42] shows higher available area for combustion under O_2/CO_2 atmospheres, *i.e.*, the pore structure.

The following issues are then under investigation and mean the main differences between fuel particle combustion under oxy-fuel and air modes:

- Diffusion governs particle conversion under high CO_2 concentration atmospheres.

- Higher CO_2 heat capacity and CO_2-char reactions could explain the ignition delay.

- Although devolatilization is not affected, residual char morphology is different.

- Higher O_2 concentration compensates the lower O_2 diffusivity in CO_2 atmospheres.

2.2. Combustion in a FB

Once the main aspects of the particle conversion are explained, we can now address the combustion performance under the particular conditions of oxy-fuel fluidized bed combustors. The O_2 concentration at the inlet, the excess of O_2 and the fluidization velocity greatly influences the combustion efficiency in fluidized bed.

Experimental operation in pilot plants allows wider flexibility, with O$_2$ at inlet ranging between 10% and 60%. Fig. **3** represents how an increase in O$_2$ concentration affects the combustion inside the fluidized bed for fixed boiler geometry.

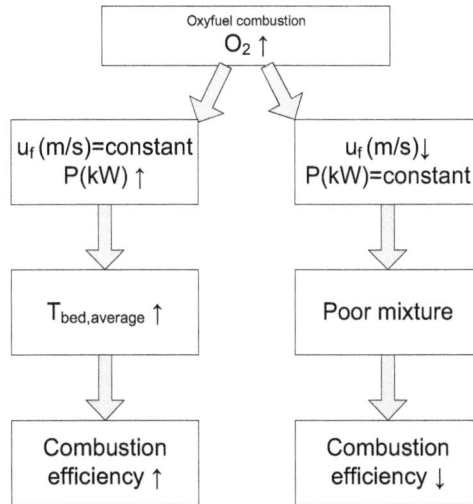

Figure 3: Effect of increasing O$_2$ concentration on the combustion efficiency.

The consequences of increasing the O$_2$ concentration also depend on the type of fuel. For example, when a low volatile fuel is fed in the fluidized bed, the combustion takes place mainly in the dense zone and the higher oxygen concentration increases temperature in this area. On the other hand, high-volatile fuels promote more uniform temperature distribution along the boiler, since a fraction of the volatiles is burnt in the diluted zone.

2.3. Temperature

Temperature profile in a fluidized bed depends on the O$_2$ concentration and fuel properties. Fig. **4** shows temperature profile in the oxy-fuel CIRCE BFB. Higher O$_2$ concentration involves higher average bed temperature. The reason is twofold: on the one hand, the adiabatic flame temperature increases with O$_2$ concentration; on the other hand, the stoichiometric fuel is highly related to the O$_2$ content in the fluidizing gas, when concentration is incremented. Thus, fuel combustion energy per unit of boiler wall surface increases, and the volumetric flow of gases decreases, diminishing the bed cooling in the diluting zone.

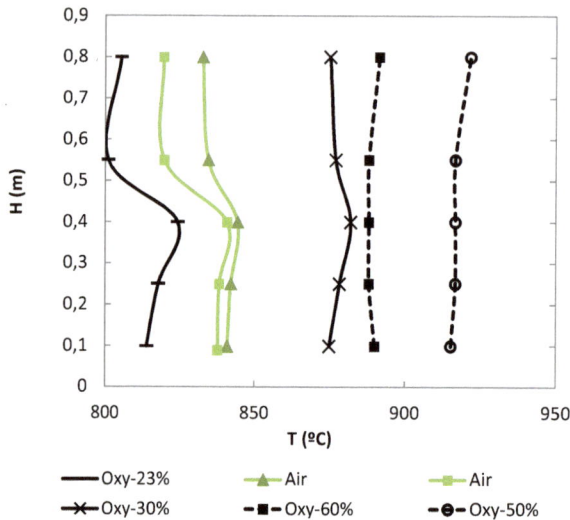

Figure 4: Temperature profile of CIRCE´s pilot plant for different fluidizing gas compositions and anthracite as fuel.

Temperature profiles in Fig. **4** reveals more uniform profiles for higher O$_2$ at the inlet. Similar profiles are obtained in other pilot plants [43, 44].

2.4. CO Emissions

Temperatures, particle size, volatile matter of fuel and fluidization velocity affect CO formation also during conventional air-firing combustion. The new O$_2$/CO$_2$ atmosphere influences the combustion performance in the way predicted by the particle conversion experience:

- High CO$_2$ concentration enhances char gasification, see Eq. (1):

$$C + CO_2 \rightarrow 2\ CO...\tag{1}$$

- Boundary layer oxidation of CO in FBC affects the mass transfer of O$_2$ to the particle surface.

- High O$_2$ concentration promotes higher burning rate and particle temperature, and thus lower CO formation.

- O$_2$ diffusivity lowers, involving higher CO formation.

Numerous factors influence the CO formation and it is not straightforward to conclude which one is the predominant.

Combustion experiences in fluidized bed found lower CO emissions under O_2/CO_2 [43, 44]. Two main factors could promote CO oxidation: higher O_2 enhances fuel reactivity, and higher thermal fuel inputs leads to higher bed temperature. Thus, there is not clear agreement if the CO decrease is dominated by one or other phenomena. Furthermore, in pilot plants, with fixed boiler size, O_2/CO_2 mixture changes the gas residence time, *i.e.* the higher the O_2 concentration, the lower the gas flow rate and the lower the fluidization velocity. On the other hand, to keep constant fluidizing velocity, fuel input must increase and, again, so it does the bed temperature.

Fig. **5** shows CO emissions measured during CIRCE BFB operation. It displays the influence of three different atmosphere compositions on the CO formation ratio (defined as $CO/(CO_2+CO)$ [22]). The secondary axle indicates the bed temperature at the dense zone (square symbols) and diluted zone (triangle symbols). The highest CO formation ratio was obtained during air-firing operation. The dilution of O_2 in RFG, instead of pure CO_2 from bottles also increases CO emissions. The main cause is the new reaction Eq. (2) with the water present in the RFG:

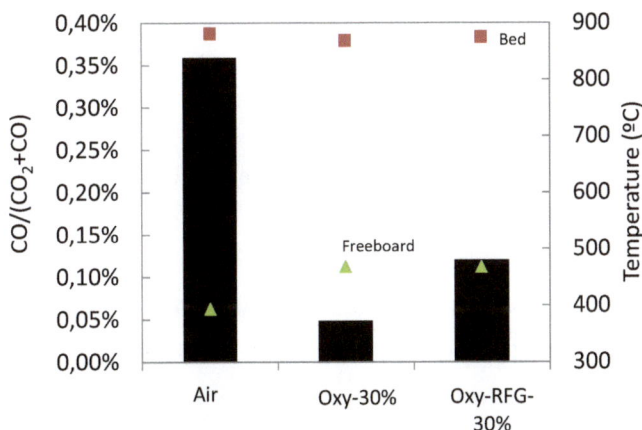

$$C + H_2O \rightarrow CO + H_2 \ldots \tag{2}$$

Figure 5: CO emissions of bituminous coal in CIRCE's facility.

Similar ratios of CO formation were obtained by Czakiert *et al.,* [45] under oxy-firing conditions. Investigations conducted on a 0.3 MW$_{th}$ CFB [46] also concluded that increasing the oxygen concentration decreases the CO emissions. Still, as in CIRCE BFB, the higher bed temperature could be the predominant factor, rather than the gas composition itself.

The influence of fluidizing velocity on CO formation has been experimentally studied by several authors [47, 48]. Whereas high velocity promotes better contact between fuel particles and gases, it also diminishes particles residence time, and so, it hinders proper contact between gas and fuel particles. Some authors [49-51] found higher CO in the exhaust gases when velocity increases. In CIRCE BFB, the increase of the residence time caused by lower fluidizing velocity in oxy-fuel operation, dominates the CO emissions (Fig. **6**). Higher velocities correspond to air cases. Velocity close to 1 m/s reaches optimum CO ratio, corresponding to oxy-firing operation.

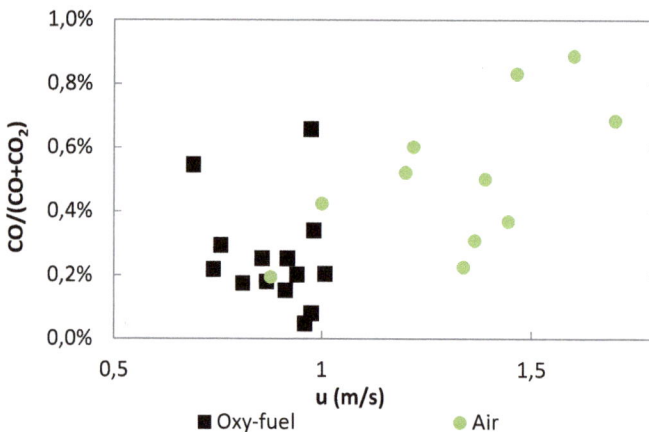

Figure 6: CO emissions *vs.* fluidization velocity for anthracite in CIRCE's facility.

2.5. Combustion Efficiency

O$_2$ concentration, excess of oxygen and fluidization velocity, determine the combustion efficiency in a fluidized bed boiler. The combustion efficiency is defined in Eq (3):

$$\eta_{comb} = 1 - \frac{C_{unburnt}}{C_{in}} \ldots \tag{3}$$

The term $C_{unburnt}$ includes the CO emissions and the unburnt elutriated fuel particles. During conventional combustion, higher temperature and higher volatile content fuel decreases unburnt carbon in the boilers, as explained by Oka [48]. Oxy-fuel follows similar tendencies. However, higher bed temperature is often related, as previously mentioned, to a higher O_2 concentration.

Fluidization velocity influence on combustion efficiency is again a trade-off between two effects. Higher velocity implies a higher flow rate of elutriated unburnt particles and, on the other hand, low velocities involve a poor mixing between gas and solids. Obviously the fuel reactivity and the fragmentation of the fuel are also crucial to determinate the influence of the velocity.

In CIRCE´S facility, the global combustion efficiency increases with the excess of oxygen and with a higher O_2 concentration. Thermal power input is higher and thus, average bed temperature also increases, as observed in Fig. **7b**.

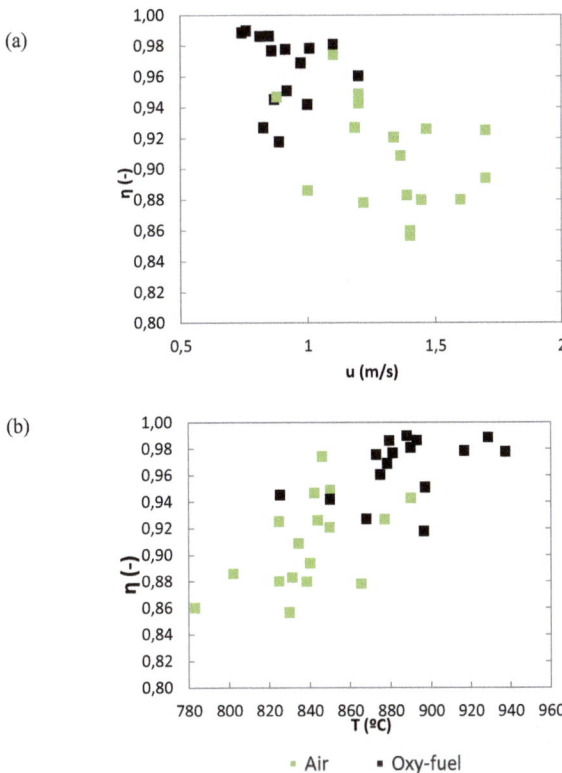

Figure 7: Efficiency combustion *vs.* fluidization velocity (a) and bed temperature (b).

Combustion efficiency using anthracite ranged between 84% and 98%. The optimum value of fluidization velocity was near 1 m/s, Fig. **7a**. In oxy-fuel operation, the higher bed temperature, together with the lower fluidization velocity and better solids back-mixing resulted in average combustion efficiency of 98%, whereas air-firing case efficiency was near 92%.

As it has been shown in this previous section, O$_2$ at inlet improves combustion efficiency. However, this could leads to lower purity of CO$_2$ in the flue gases. When operating at higher O$_2$ concentration, circulating gases diminish, and the excess of oxygen and other impurities reach higher percentages in the output current stream. This will be a relevant issue when comparing pollutant emissions between air and oxy-fuel, as will be later explored.

3. SO$_2$ EMISSIONS

Sulphur content in fossil fuels can range between 0% and 10%. A major fraction of sulphur is released during fuel pyrolysis and its distribution among volatile matter, char and tar depends on the coal rank and combustion conditions. The sulphur released during devolatilization is proportional to the volatile matter content of fuel, so the content of sulphur in char is higher in high rank coals like anthracite.

Fluidized bed combustion takes advantage of the variety of particles comprising the bed to reduce SO$_2$. The presence of coal ashes, containing alkali metals as Ca or Mg, helps to retain SO$_2$. This is the so-called auto-retention, since SO$_2$ is captured by burnt coal particles themselves. This is common in low rank coals with high ash content in their composition.

In most cases, auto-retention is not enough to reduce SO$_2$ emission down to the values required by environmental regulations, and additional SO$_2$ retention methods are required. The addition of a flue gas desulphurization unit, downstream, is the common option for pulverized fuel boilers. In fluidized bed boilers, however, *in situ* desulphurization is much more advantageous, by the addition of high Ca or Mg content sorbents, like limestone or dolomite. Limestone is the most frequently used sorbent because of availability and cost, whereas dolomite is recommended for pressurized fluidized bed due to the properties of $MgCO_3$ to capture SO_2 at high pressure.

3.1. Desulphurization Mechanism

In Fig. **8**, equilibrium CO_2 partial pressure (P_{CO2}) is represented against temperature. Under conventional combustion in atmospheric fluidized beds, P_{CO2} is around 12 kPa and temperature is over 800 °C. Limestone is then involved in two reactions to capture SO_2. First, it calcines to form CaO and releases CO_2, Eq. (4). This reaction takes place over 650 °C and generates a porous particle. The second reaction is the SO_2 capture itself, obtaining $CaSO_4$ (Eq. (5)).

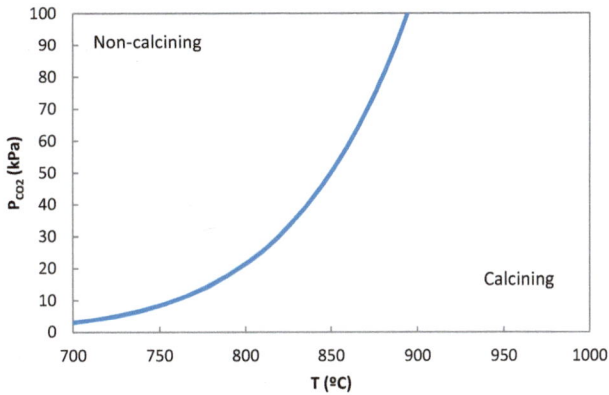

$$CaCO_3 \leftrightarrow CaO + CO_2 \ldots \tag{4}$$

$$CaO + 1/2\, O_2 + SO_2 \rightarrow CaSO_4 \ldots \tag{5}$$

Figure 8: Equilibrium P_{CO2} against temperature.

The desulphurization mechanism changes when P_{CO2} is increased, like in pressurized fluidized bed or in oxy-fuel combustion. High P_{CO2} inhibits CO_2 migration out of the sorbent particle and calcination does not take place. Desulphurization is then carried out in one step, by means of direct sulphation (Eq. (6)).

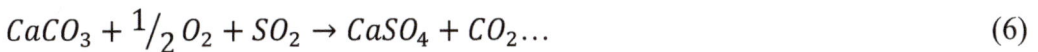

$$CaCO_3 + {}^1/_2\, O_2 + SO_2 \rightarrow CaSO_4 + CO_2 \ldots \tag{6}$$

If limestone calcines, the sorbent particle becomes a highly porous product, so SO_2 can penetrate easily into the particle and more Ca is consumed. Depending on the particle diameter and limestone properties, a sorbent particle can capture SO$_2$ by means two different modes. If particle diameter is small and CaO is porous enough, particle is sulphated uniformly and a high percentage of Ca is used. The most common model states that a CaSO$_4$ layer is developed around sorbent particle. This blocks the pores because CaSO$_4$ specific volume is higher than CaCO$_3$. Therefore, there is a considerable unreacted amount of Ca in the sorbent particle. This explains the need of introducing higher Ca over stoichiometric to optimize SO$_2$ capture.

Direct sulphation has the disadvantage of being slower than indirect capture. The counter current created by SO$_2$ entering into the particle and CO$_2$ released during sulphation, generates a more porous CaSO$_4$ layer than when indirect sulphation takes place, enhancing the use of the sorbent. Nevertheless, the use of limestone under high P$_{CO2}$ conditions has a strong dependency on particle size. The low porosity of limestone does not allow an easy SO$_2$ diffusion into the particle. This effect augments with the particle size and higher Ca cannot react. For direct sulphation small particles (<100 microns) reach values similar to those obtained during indirect sulphation, although they need longer residence times. The evolution of product layer during sulphation differs in both mechanisms. In indirect sulphation CaSO$_4$ layer develops in the first stage and then it is followed by residual activity. In direct sulphation reaction rate keeps constant for a longer period [52].

Table **3** summarized the most relevant findings of SO$_2$ capture under oxy-firing conditions.

An important difference between the behaviors of particle sorbent in air and oxy-firing is related to the fragmentation and the formation of smaller particles. Fragmentation is caused by thermal shock, internal overpressures and impacts against walls and other particles. These smaller particles create new available surfaces that improve desulphurization. On the other hand, these new particles elutriate easier without interact with SO$_2$, with negative effects on SO$_2$ capture efficiencies. Particle fragmentation is reduced when direct sulphation happens [53].

3.2. Influence of Operation Parameters on Desulphurization

Bed temperature is probably the most influent parameter together with P$_{CO2}$ to optimize desulphurization efficiency. Under conventional combustion conditions, the optimum temperature ranges from 825 °C to 850 °C. Below this temperature, desulphurization reaction becomes slower and over it, reductive reactions affect CaSO$_4$, releasing SO$_2$ previously captured. It results then in a competition between desulphurization and reduction reactions [54].

Table 3: Main research results about desulphurization under oxy-firing conditions

Group	Reactor	Fuel	Fluidizing Gas	Conclusion
INCAR [55, 56]	TGA	-	Pressurized Air	High CO$_2$ partial pressure changes the desulphurization mechanism to direct sulphatation. Direct sulphatation is strongly controlled by intraparticle diffusivity.
CANMET [20, 57, 58]	CFB	Bit. coal Petcoke	Air 23/77 O$_2$/RFG 30/70 O$_2$/RFG 35/65-45/55 O$_2$/RFG	Desulphurization efficiency decreases under OF conditions. SO$_2$ concentration in OF increases up to 5 times against AF. Indirect sulphation increases desulphurization efficiency for petcoke.
Czestochowa University of Technology [22, 45]	CFB	Bit. coal	Enriched air (35%) 21/79-35/65 O$_2$/CO$_2$	S-fuel conversion increases in OF.
Southeast University of Nanjing [59, 60]	CFB	Anthracite Bit. coal	21/79-40/60 O$_2$/CO$_2$	SO$_2$ increases with O$_2$ concentration, as well as desulphurization efficiency.
IRC-CNR [61, 62]	BFB	-	0/100-40/60 O$_2$/CO$_2$	Primary fragmentation of limestone under oxy-firing was less intensive than in air-firing. Attrition during oxy-firing is reduced favoring lower Ca losses.
CSIC-ICB [24, 63-65]	TGA BFB	Anthracite Lignite	Air 21/79-40/60 O$_2$/CO$_2$	Limestone conversion lowers under non-calcining conditions. Optimum desulphurization temperature was found at 925 °C, under calcining conditions.
CIRCE [66]	BFB	Bit. Coal Lignite	Air 23/77-60/40 O$_2$/CO$_2$	There is no optimum temperature for desulphurization under 900 °C. High desulphurization efficiency can be reach under oxy-firing conditions but it is lower than in conventional combustion.

Fig. **9** shows results obtained from combustion of lignite in CIRCE BFB. The optimum temperature under conventional combustion conditions is found at 850 °C (black symbols).

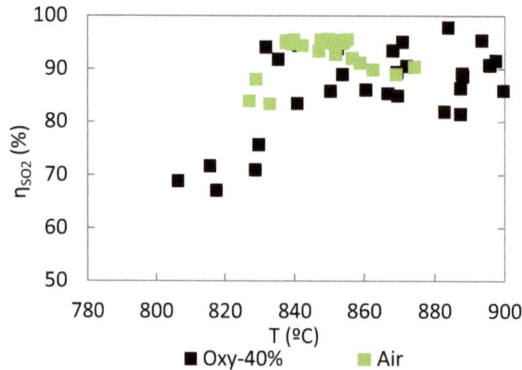

Figure 9: Desulphurization efficiency against temperature under air and oxy-firing of lignite.

Under oxy-firing conditions (brighter symbols), the desulphurization optimum temperature is not perceived in this operational range. With CO$_2$ concentration around 60% and bed temperature below 900 °C, direct sulphation takes place. Several experiments carried out in CANMET CFB testing different limestones and fuels reported lower desulphurization efficiency when direct sulphation took place [20, 57, 58]. Combustion tests with lignite and a bituminous coal resulted in an increase of desulphurization efficiency when increasing bed temperature. For temperature higher than 900 °C indirect sulphation begins to dominate desulphurization. Then, SO$_2$ capture efficiency increases, but usually at lower levels than those obtained under air-firing. Optimum desulphurization temperature under oxy-firing conditions was found around 925 °C [64].

Another option to improve desulphurization efficiency is increasing Ca:S ratio. This way, Ca available in the reactor can compensate the low kinetic rate of direct capture and the lower use of Ca.

The availability of O$_2$ has no influence on desulphurization kinetic rate when it is over 4% and some researchers neglect this effect. However, O$_2$ concentration affects S-fuel conversion into SO$_2$. Czakiert *et al.,* [22] found that conversion is increased at O$_2$ concentration up to 30% and over this percentage it decreases due to the auto-retention capture.

Another parameter that affects desulphurization is particle diameter. As it is said above, the use of limestone particle is drastically reduced under direct sulphation conditions when the diameter is increased. If indirect sulphation conditions take place, it was observed a lower Ca conversion of large particles since CaSO$_4$ layer is developed around the core of the particle avoiding CaO core sulphation.

4. NO$_X$ EMISSIONS

N$_2$, NO and NO$_2$ are usually the main forms in which N is contained in combustion flue gas. Moreover, in fluidized beds, the presence of N$_2$O is considerable promoted by the lower temperature.

NO$_x$ chemistry is a complex mechanism of formation and reduction reactions in which a vast number of substances are involved, as shown in Fig. **10**. The source of NO$_x$ from combustion can be either fuel-N or N$_2$ contained in the oxidizing gas. Low operation temperature of fluidized bed avoids thermal- NO$_x$ mechanism, so NO formed from the latter source is prevented. During oxy-fuel, the N present in oxidizing gas can be considered negligible and the only source of N considered is fuel.

The release of fuel-N depends on the particle heating rate and the temperature. High temperature increases volatile matter release and thus, the N contained in gas phase. The ratio between volatile-N and char-N is highly influenced by both, heating rate and temperature. This ratio determines the NO precursors and where they are formed along the boiler.

Volatile-N is mainly released as HCN and NH$_3$. The former is the main precursor of N$_2$O, whereas the second tends to form NO. N$_2$O is unstable at temperatures over 800 °C. At usual fluidized bed temperatures a net N$_2$O destruction is obtained, although considerable concentration remains in exhaust gases. Over 800 °C, free radicals concentration increases and enhances N$_2$O reactions to form NO or N$_2$. HCN also forms NH$_3$, which is a precursor of NO.

Gaseous NO precursors are also formed from char-N, mainly HCN, following the same mechanism as the volatile-N compounds. Heterogeneous reactions takes place between fuel particles and gaseous compounds generating NO, that reacts with other available compounds. The interactions among NO precursors and solid particles influence the NO$_x$ formation. For example, CaO presence changes the selectivity of HCN and it reacts to form NO rather than N$_2$O. NO can interact with other substances present in the reactor, like CO or NO, and it can be destroyed by reduction.

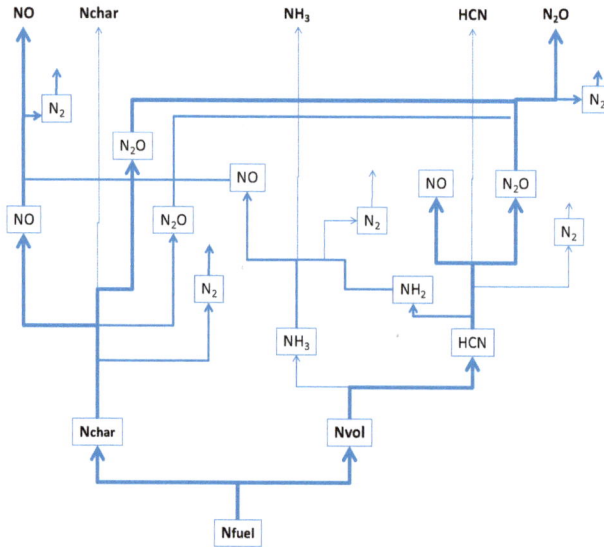

Figure 10: Main conversion paths of fuel-N. Adapted from [67].

During conventional combustion temperature, coal rank, fuel particle diameter and excess of O$_2$ are the main parameters influencing NO emission:

- Increasing temperature improves combustion, so more char-N releases. Also, higher free radical concentration increases, thus NO formation is favored. CO concentration diminishes and limits the possibilities of reducing NO by means reaction with char, CO and NO.

- The influence of coal rank is related to volatile-N and char-N distribution and how the particles embed nitrogen. HCN is more abundant in high rank coals combustion but HCN usually forms N$_2$O instead of NO. Low rank coals released more NH$_3$ which forms NO. This way, fuel-N conversion into NO is higher for low rank coals.

- Particle size affects N released with volatile matter. Heating rate in coarser particles lowers, and diminishes volatile-N release.

- NO is formed by oxidation, so O$_2$ availability greatly influences emissions. Typically, O$_2$ excess over stoichiometric assures proper combustion efficiency, but it enhances fuel-N conversion. In oxy-firing the O$_2$ availability is determined not only by the excess over stoichiometric, but also by the concentration of O$_2$ in the oxidizing stream.

Particularly in oxy-firing, the presence of high CO$_2$ concentration promotes gasification reactions, increasing the concentrations of CO and H, and thus, modifying the NO formation and reduction mechanism.

Secondary air injection is an interesting technique for further reduce NO emissions. It consists on the injection of part of the oxidizing gas at different height. This way, O$_2$ availability is modified along the reactor. In the lower part of the furnace, it is created a reductive atmosphere favoring low NO generation and the injection of secondary air completes fuel combustion. In oxy-firing, the secondary gas offers analogous advantages. In both cases, global excess of O$_2$ is not affected by the addition of secondary gas but by the share of O$_2$ availability.

When O$_2$ in oxy-firing is diluted in RFG, NO goes back to the reactor and takes part of the pool of reacting substances. This way, reduction reactions favor a decrease of emitted NO concentration.

Although experiments showed that NO emissions concentration during oxy-firing conditions is higher than in conventional combustion, the main cause lies on the lower gas volume. NO emissions represented per energy units avoid any influence of gas volume and values of NO emission are usually similar for both atmospheres. Fig. **11** shows the NO$_x$ measurements collected during CIRCE BFB operation, represented in both units against bed temperature.

An increase of temperature has negative effect on NO emission. During the tests, this increase coincides with higher combustion efficiency, and then, less CO present in gases, decreasing the NO reduction. This was so when burning high rank coal (bituminous). However, lignite combustion changed this tendency and no clear influence of temperature was observed. This confirms that coal rank directly affects combustion and fuel-N release. Fuel-N conversion for oxy-fuel combustion of lignite reached values up to 30%, whereas bituminous fuel-N conversions ranged between 5% and 20%. The higher fraction of volatile matter in low rank coals favors total N released during the first stage of combustion. It forms NO in higher zones of the reactor and, consequently, it is hardly reduced to N$_2$.

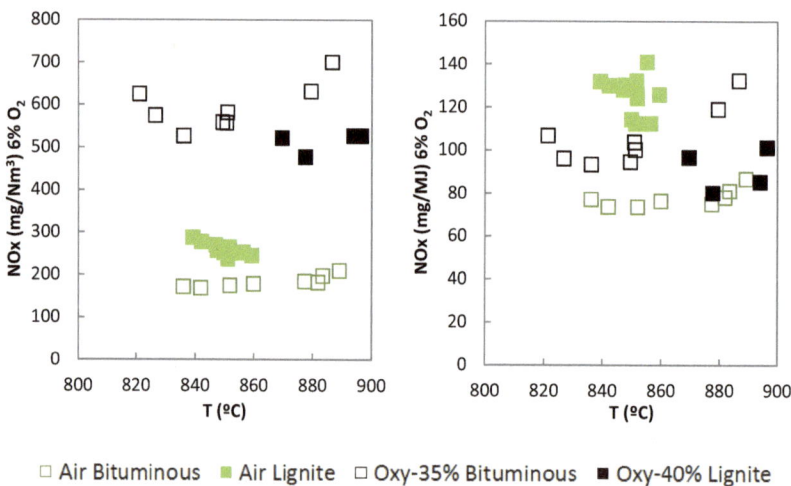

☐ Air Bituminous   ■ Air Lignite   ☐ Oxy-35% Bituminous   ■ Oxy-40% Lignite

Figure 11: NO emission generated under air-firing and oxy-firing conditions.

Fig. **12** shows the significant effect of O$_2$ excess on NO emission. During the four experiments shown, NO clearly increase with excess of O$_2$. This is caused by the better fuel particle conversion which implies a considerable increase of fuel-N conversion, together a lower CO concentration.

Experiments of oxy-fuel combustion with RFG revealed that fuel-N conversion reduced from 6% under air-firing to 1% under oxy-firing with 60% RFG [20, 57, 58]. Recirculation of NO in flue gas, allows its reduction by char particles.

Duan and co-workers [68] explored the effect of gas-staging on NO emissions. They reported a decrease of NO emission when primary gas was reduced. This effect was more important for higher O$_2$ concentrations. They tested the possibility of introducing different O$_2$ concentration in primary and secondary inlets. Increasing O$_2$ concentration contained in primary gas significantly incremented NO emission, since the effect of the reducing ambient is offset. Coal rank influence was explored in CIRCE BFB. Low rank coals, as lignite, with a higher volatile content, suffers a lower reduction of NO emission by means of gas-staging than high rank coals. An important fraction of fuel-N is released during devolatilization of low rank coals. Thus, the possibility that NO precursors interact with bed particles or other compounds drops, reaching higher fuel-N conversion.

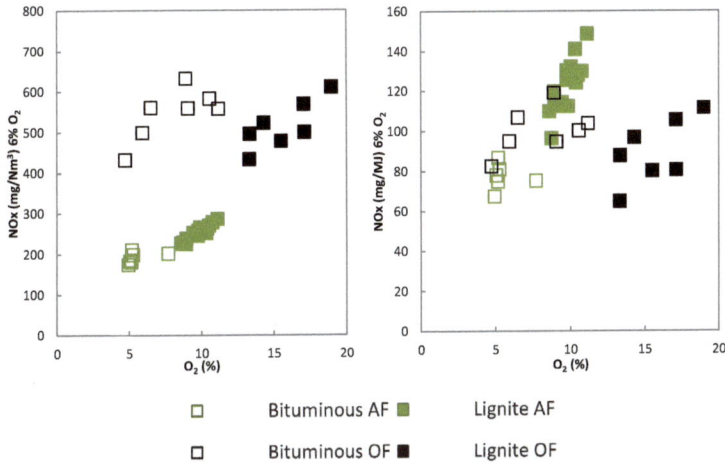

Figure 12: Influence of excess of O$_2$ on NO emission under air-firing and oxy-firing conditions of two different coals.

Table **4** shows some of the most important findings related with NO emitted during oxy-firing in a fluidized bed.

Table 4: Main findings about NO emissions during oxy-firing in a fluidized bed

Authors	Reactor	Fuel	Fluidizing Gas	Conclusion
CANMET [20, 57, 58]	CFB	Bit. coal Petcoke	Air 23/77 O$_2$/RFG 30/70 O$_2$/RFG 35/65-45/55 O$_2$/RFG	NO$_x$ emission decreases under OF even at higher temperatures. Extremely low fuel-N conversion under OF was observed.
Czestochowa University of Technology [22, 45]	CFB	Bit. coal	Enriched air (35%) 21/79-35/65 O$_2$/CO$_2$	OF favors fuel-N conversion due to higher O$_2$ partial pressure.

Table 4: contd….

| Southeast University of Nanjing [60, 68, 69] | Gas tight tube furnace CFB | Anthracite Bit. coal | Ar-CO$_2$ 21/79-40/60 O$_2$/CO$_2$ | CO$_2$ suppresses NH$_3$ formation and increases HCN. N conversion is less intense at low temperatures in CO$_2$ atmospheres but it is more important at high temperatures, in comparison with conventional combustion atmospheres. NO increases with O$_2$ concentration. |
| CIRCE [66] | BFB | Bit. Coal Lignite | Air 23/77-60/40 O$_2$/CO$_2$ | NO$_x$ emissions per energy unit are similar for air-firing and oxy-firing. NO is highly influenced by excess O$_2$. |

5. ENERGY BALANCE IN OXY-FUEL FLUIDIZED BED BOILERS

In general, the control of temperature in fluidized beds reaches particular relevance because the pollutant emissions control and the risk of particles agglomeration. In oxy-fuel combustion there are two additional issues that affects this balance, compared to the conventional air combustion: i) The reduction of gas flow allows more compact boilers, and ii) the differences on gas composition, influencing heat transfer coefficients.

5.1. Reduction of Gas Flow Rate

In general, heat transfer surfaces arrangement in a FB combustor consists of evaporative, superheating and reheating zones. In tubes bundles by the wall, evaporation usually occurs. The hot flue gas exiting the cyclone enters in the convective pass, located downstream of the combustor, and transfers heat to super heaters and economizer respectively. In some CFB arrangements, an extra heat transfer surface exists in the fluidized bed external heat exchanger (FB-EHE). There, collected solids from the cyclone cool down before recycling back to the boiler, through the sealing device. This heat transfer arrangement will differ in oxy-fuel combustion. Oxygen concentration is higher and the diluting substance is mostly CO$_2$, denser than the air-N$_2$. Less gas flow travels through the boiler and so, more compact boilers are expected. Saastamonien *et al.*, [18] calculated an available surface in a oxy-fuel boiler 38% smaller than in air-firing when O$_2$ at inlet is 60%. They estimated that higher heat flux surfaces would lead to significant reduction in the boiler size. The available area for exchange heat along the riser walls is then reduced for certain fuel input.

At the same time, the smaller flow rate of gas diminishes the heat transfer by gas convection in the convective pass. A model of the 670 t/h lignite CFB boiler in Turow power station in Poland [70] considered mixtures of O_2/CO_2 and O_2/N_2 gives a limit of 60% of O_2 at inlet, stating that bed temperature over this concentration will exceed 1050 °C and so ash melting would take place.

The possibility of accounting for extra heat transfer surfaces in the FB-EHE presents several advantages for the flexibility of oxy-fuel systems. Simulations by Nsakala *et al.,* [17] confirmed the significance of an external way of transferring heat from particles. Additionally they observed a potential problem of using an external fluidized bed as heat exchanger. If bed temperature is so high that calcination of limestone takes place, the calcium oxide arriving to the FB-EHE could react with the fluidizing gas, consisting mainly of CO_2. If re-carbonation reaction takes place, fluidization could fail, giving place also to tubes sintering problems.

Heat transfer in fluidized beds is strongly related to the bulk density at each height. If available heat transfer area diminishes, a way of enhancing heat transfer is increasing particles travelling along the riser. Seddighi *et al.,* [19] predicted particle recirculation rates higher than 30 kg/m^{-2}s^{-1} for controlling bed temperature at O_2 concentration as high as 90% at inlet.

Fluidized bed becomes then especially advantageous for oxy-fuel because heat transfer surface arrangement could be designed either for retrofitting existing boilers or for reducing boiler size, increasing O_2 concentration at inlet. Both ideas need to be addressed in the near future for improving overall oxy-fuel combustion process efficiency. Fig. **13** shows the simulation results and compares heat shares transferred in the riser walls surface (*Rate Walls*), in the convective pass (*Rate Flue Gas*) and in the FB-EHE (*Rate Loop Seal*), for different oxidant streams compositions. The cross section of the boiler is indicated and also the value of elutriated solids, Gs, in kg m^{-2}s^{-1}. Air-firing case and 30% O_2 at inlet, presents similar values for boiler cross sectional requirements. This is in agreement with the experiments that confirm a percentage of 29% O_2 at inlet for reaching similar gas conditions than air-firing case. If O_2 concentration increases, the boiler turns out more compact. In the case of 45% O_2, the cross area reduces to half from the original size. Thus, Gs increases cooling capacity in the FB-EHE around 15 points. An even higher increase in Gs values, up to 32 kg m^{-2}s^{-1} for 60% O_2, is caused by reducing cross sectional area down to 34%. Less available area in the riser for exchanging heat confirms the essential role of external heat transfer surfaces.

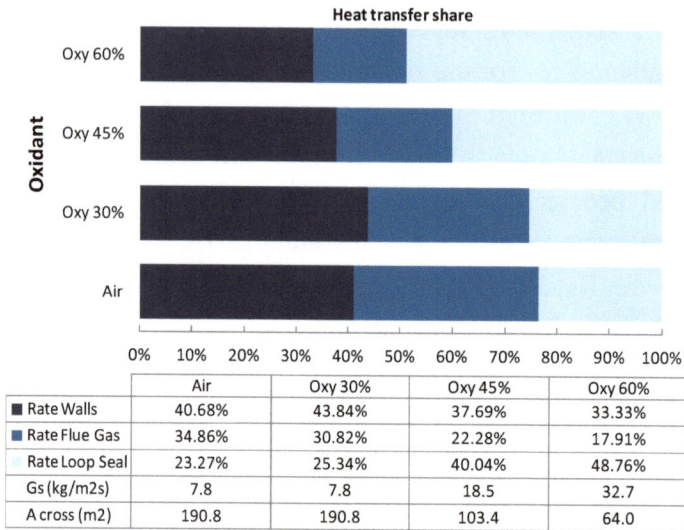

Heat transfer share

	Air	Oxy 30%	Oxy 45%	Oxy 60%
■ Rate Walls	40.68%	43.84%	37.69%	33.33%
■ Rate Flue Gas	34.86%	30.82%	22.28%	17.91%
Rate Loop Seal	23.27%	25.34%	40.04%	48.76%
Gs (kg/m2s)	7.8	7.8	18.5	32.7
A cross (m2)	190.8	190.8	103.4	64.0

Figure 13: Heat balance simulation results for same fuel input 650 MW.

If fuel input increases, an adequate temperature profile along the furnace is possible for higher O_2 concentrations. In this way, higher volume of gases enters into the boiler (for stoichiometry) and so, higher velocity entrains more solids to the cyclone.

Fig. **14** shows the distribution heat transfer share, for keeping similar temperature profile in the boiler, but different composition of the oxidant stream. In the case of 40% O_2 at inlet two possibilities allows an adequate heat transfer share in the external heat exchanger: increasing elutriated solids, by fuel input, and so, the primary gas velocity; or increasing the cooling of recycled solids in the FB-EHE. This last option will required a considerable increment of heat transfer surface and a re-arrangement of conventional FB-EHE, for fitting the temperature requirements. In the former possibility, attention must be paid to a proper fluidization of solids in the FB-EHE. Higher solids rate arriving from the cyclone must be conveyed along to the recycling pipe to the CFB. This involves the risk of poor fluidization points and so, the lower heat transfer at certain points of the EHE.

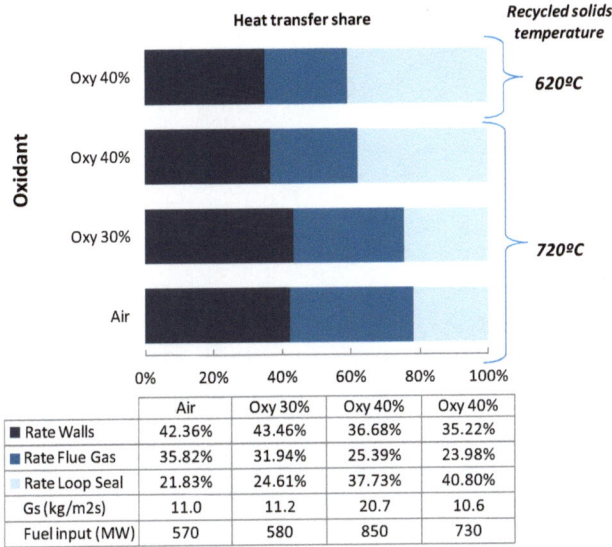

	Air	Oxy 30%	Oxy 40%	Oxy 40%
■ Rate Walls	42.36%	43.46%	36.68%	35.22%
■ Rate Flue Gas	35.82%	31.94%	25.39%	23.98%
Rate Loop Seal	21.83%	24.61%	37.73%	40.80%
Gs (kg/m2s)	11.0	11.2	20.7	10.6
Fuel input (MW)	570	580	850	730

Figure 14: Heat balance simulation results for air boiler size and different O_2 concentrations at inlet.

The FB-EHE works under bubbling conditions and it is fluidized with air, favouring also the combustion of unburnt particles in this device. In oxy-firing case, air cannot fluidize the FB-EHE and other gases will be used, such as recycled flue gas. The influence of fluidization gas composition on the heat transfer in a bubbling combustor is following explore, through the experimental results in the CIRCE pilot plant.

5.2. Heat Transfer Coefficient in Oxy-Fuel Fluidized Beds

Heat transfer in fluidized bed is usually conceived as the independent contribution of three mechanisms, see Eq (7): the gaseous convection (h_{gc}) of bubbles; particles convection (h_{pc}) due to transient conduction of the emulsion phase; and thermal radiation (h_{rad}) due to high temperature:

$$h = h_{gc} + h_{pc} + h_{rad} \cdots \tag{7}$$

In oxy-fuel, gas composition greatly differs from that in air-firing. Fig. **15** represents gas density, viscosity, conductivity and specific heat capacity for different O_2/CO_2 mixtures.

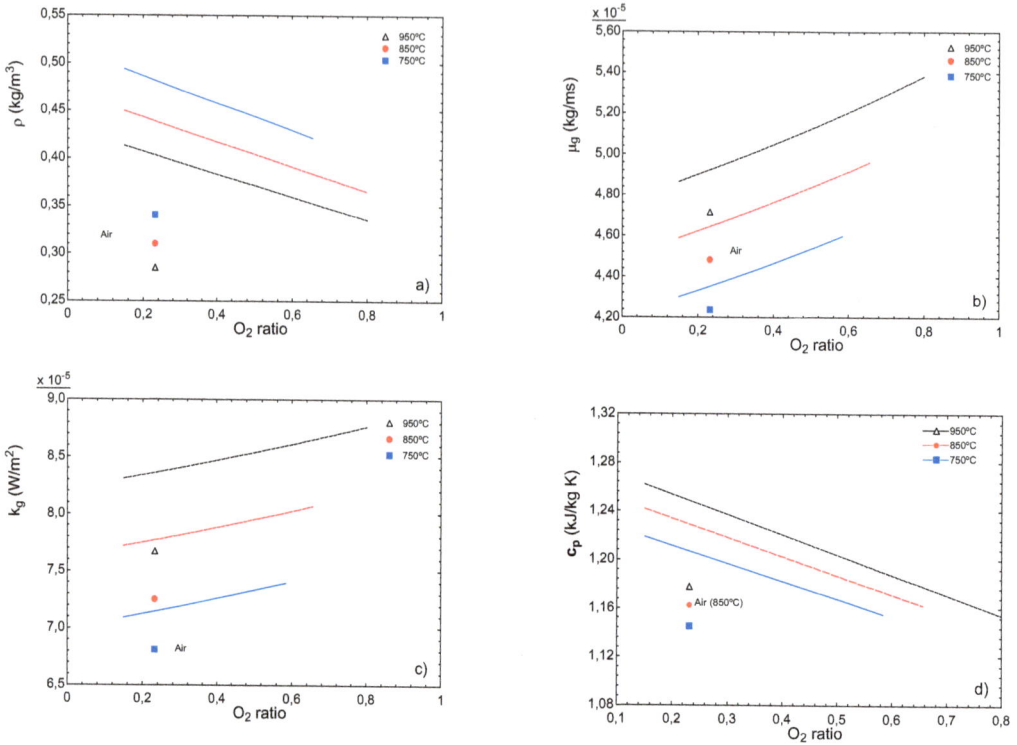

Figure 15: Gas properties variation for different O$_2$/CO$_2$ mixtures at inlet: a) density; b) viscosity; c) conductivity; d) specific heat capacity.

Gas properties differ substantially between an atmosphere of oxygen with nitrogen and an atmosphere of oxygen with carbon dioxide. The viscosity, Fig. **15** b), is calculated by Wilke correlation [71]. The influence of this parameter is closely related also to fluidization, since it represents the resistance to the flux of particles in the fluid. Thermal properties of gas, such as conductivity and heating value, differ from air to oxy-firing as well. In FB, the solids thermal properties are much higher than gas and so, these are limiting factor of heat transfer.

During oxy-fuel combustion, the concentration of tri-atomic gas molecules in the flue gas increases drastically and will change the emissivity of the gas. The contribution of radiation heat transfer mechanism to the overall heat transfer in fluidized beds is not as critical as in PC boilers. In the diluted part of a fluidized bed, the radiation and convection dominates the heat transfer, in different ratios, but in bubbling fluidized bed, the contribution represents around 20% of the total heat transfer coefficient. The radiation contribution is very much dependant on the shadow effect of particles falling by the riser walls [72]. In the denser zones, these mechanisms loose relevance and particle convection prevails.

The dense zone (the lower part of the riser) in a CFB is analogous to a BFB in terms of fluid-dynamics behavior. The bubbling conditions are also common in FB-EHE devices. In this fluid-dynamics regime, the heat transfer caused by the movement of particles dominates over gas convection and radiation.

In a BFB, the contribution of the heat transfer by gas convection takes place while bubbles are contacting the surface. This gas convection mechanism is one order of magnitude lower than the heat transfer by the solids movement. This is the reason why values obtained during oxy-fuel are not expected to differ considerably from those obtained during air-firing. During the time that the particles contact the wall, conduction heat transfer takes place through the small contact point of the particle with the wall but mostly, through the gas film adjacent between particle and wall (Fig. **16**). This is the way in which gas composition would influence heat transfer by particle convection.

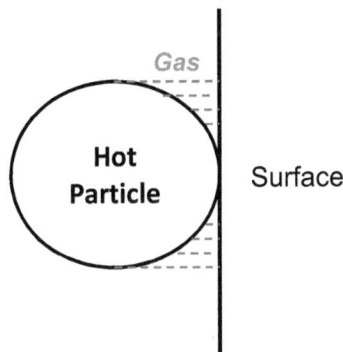

Figure 16: Heat transfer by particle convection. Adapted from [73].

During the operation of the CIRCE BFB pilot plant, heat transfer was measured by energy balances in the cooling jackets, aiming to find any effect of gas properties changes during oxy-firing. Fig. **17a** shows the heat transfer coefficients measured during the different tests for controlling the bed temperature. The horizontal axle represents the concentration of O$_2$ at inlet in every case.

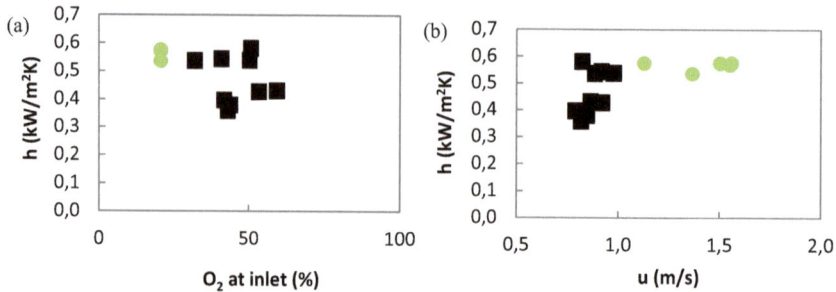

Figure 17: Heat transfer coefficients measured in the oxy-fuel BFB in CIRCE *versus* a) O$_2$ concentrations at inlet, b) fluidizing velocity.

The ascendant tendency on gas conductivity theoretically observed in Fig. **15c** does not show a relevant effect on heat transfer coefficient. The reason is that the dominant parameter determining heat transfer coefficients is the fluidizing velocity, as observed in Fig. **17b**. For the same boiler design, increasing O$_2$ concentration means less mass flow required for combustion and also, the presence of CO$_2$ instead of N$_2$, makes the flow denser. This fact, added to the higher fuel input due to higher O$_2$ concentration in the oxidant stream, leads to lower velocities in the oxy-fuel tests. The increase of fluidizing velocity promotes heat transfer due to the higher renewal frequency of "warm" particles in the wall. At certain point, further increase in fluidizing velocity diminishes heat transfer, because gas contact the wall more frequently and gas convection heat transfer is much less intense. In Fig. **17b**, oxy-fuel combustion range of operation, heat transfer increases with fluidizing velocity, whereas, for gas velocities in the air-firing case, this influence is not relevant anymore.

In future oxy-fuel FB designs the external heat exchanger surfaces will play an essential role for the proper control of bed temperatures. The new EHE will manage higher amount of solids with consequences in the fluidization uniformity. Heat transfer and fluid-dynamics are issues still under research for the future EHE designs.

CONCLUSION

Oxy-fuel combustion in a fluidized bed is being one of the most relevant technologies for the deployment of CO$_2$ capture in fossil fuel-based combustion facilities. The change of the composition of the gases used as oxidizer, with a high

CO$_2$ concentration, will modify the performance of the boiler island in many aspects.

Combustion is affected by the lower diffusivity of O$_2$ in high concentration CO$_2$ atmospheres. This produces a delay in the ignition and lower adiabatic temperature due to the higher heat capacity of the CO$_2$. The presence of high CO$_2$ concentration also enhances gasification reactions, more important if RFG contains water (wet RFG). Nevertheless, the possibility of increasing O$_2$ concentration in the oxidizer, together with the expected higher bed temperature, can reduce the emission of CO.

Desulphurization mechanism is strongly influenced by the bed temperature and CO$_2$ partial pressure inside the boiler. At typical fluidized bed temperatures, the SO$_2$ capture mechanism is carried out without previous calcination of the limestone. This process is slower and typically lower efficiencies are obtained.

NO$_x$ emissions are negatively affected by high bed temperature and excess O$_2$, generally taking place during oxy-fuel combustion. However, gasification reactions, which take place in the reactor, enhance the reduction NO by means reactions with CO and char. The use of RFG also favors the decrease of both pollutants.

Boiler energy balance will be altered by the lower gas flow. Increasing O$_2$ concentration in the oxidizer gas stream allows the decrease of the size of the boiler, limiting the heat transfer to the riser wall surface and in convective pass. The increase of O$_2$ concentration in fluidizing gas enhanced the necessity of an FB-EHE to remove part of the combustion heat form the recycled particles in the hot loop. This device will provide flexibility for the operation of the oxy-firing systems. Heat transfer coefficient of the dense zone of BFB, CFB or FB-EHE in oxy-firing is highly dependent on particle convection, being dominated by fluidization velocity.

ACKNOWLEDGEMENTS

The work described in this chapter was partially funded by the R+D Spanish National Program from the Spanish Ministry of Science and Innovation (MICINN, *Ministerio de Ciencia e Innovación*) under projects ENE-2009-08246 and ENE-2012-39114-C02-01.

CONFLICT OF INTEREST

The authors confirm that this chapter contents have no conflict of interest.

NOMENCLATURE

C (-)	Carbon
C_p (kJkg^{-1}K^{-1})	Specific heat
G_s (kg m^{-2}s^{-1})	Elutriated solids
H (m)	height
H (W m^{-2}K^{-1})	Total heat transfer coefficient
h_{gc} (kW m^{-2}K^{-1})	Gaseous convection coefficient
h_{pc} (kW m^{-2}K^{-1})	Particle conduction coefficient
h_{rad} (kW m^{-2}K^{-1})	Radiation coefficient
k_g ((kW m^{-1}K^{-1})	Gas conductivity
P (kPa)	Pressure
T (K)	Temperature
u (m/s)	Fluidzation velocity

GREEK'S SYMBOLS

ρ (kg m-^{3})	Density
μ (kg m^{-1}s^{-1})	Viscosity
η	Efficiency

ACRONYMS

AF	Air-firing
BFB	Bubbling fluidized bed
CCS	Carbon Capture and Storage
CFB	Circulating Fluidized Bed
EHE	External Heat Exchanger
FB	Fluidized Bed
NG	Natural Gas
PC	Pulverized coal
TGA	Thermogravimetric analysis

REFERENCES

[1]   Halder, S. An Experimental Perspective on Praxair's Hot Oxygen Technology to Enhance Pulverized Solid Fuel Combustion for Ironmaking Blast Furnaces, Indianapolis, USA, 2-5 May 2011; The Iron & Steel Technology Conference and Exposition; Eds.: Indianapolis, USA, **2011**.

[2]   Kobayashi, H. Advances in Oxy-Fuel Fired Glass Melting Technology: XX International Congress on Glass; Kyoto, Japan, 26 September – 1 October 2004; ed.: Kyoto, Japan, **2004**.

[3]   Lanyi, M.D. Discussion on steel burning in oxygen (from a steelmaking metallurgist's perspective) In:Flammability and Sensitivity of Materials in Oxygen-Enriched Atmospheres. Vol. 9; Steinberg, T.A.; Beeson, H.D.; Newton, B.E.; Eds. West Conshohocken, PA: American Society for Testing and Materials; **2000**, pp. 163-178

[4]   Kiga, T.; Takano, S.; Kimura, N.; Omata, K.; Okawa, M.; Mori, T; Kato, M. Characteristics of pulverized-coal combustion in the system of oxygen/recycled flue gas combustion. Energ Convers Manage, **1997**, 38, S129-S134.

[5]   Santos, S.; Haines, M. Oxy-fuel Combustion: Progress and Remaining Issues., In: 2nd Workshop of the International Oxy-Combustion Research Network., Windsor CT, USA, Eds.: Windsor CT, USA, **2007**.

[6]   IEA.Power generation from coal. Measuring and reporting efficiency Performance and CO$_2$ emissions, **2010**.

[7]   Koornneef, J.; Junginger, M.; Faaij, A. Development of fluidized bed combustion--An overview of trends, performance and cost. Prog Energ Combust, **2007**, Vol. 33, pp.19-55.

[8]   Burchhardt, U., Experiences from Commissioning and Test Operation of Vattenfall's Oxyfuel Pilot Plant, In: 1st Oxyfuel Combustion Conference; Cottbus, Germany; 8-11 September 2009; Eds.: Cottbus, Germany, **2009**.

[9]   McCauley, K.J.; Farzan, H.; Alexander, K.C.; McDonald, D.K.; Varagani, R.; Prabhakar, R.; Tranier, J.-P.; Perrin, N. Commercialization of Oxy-Coal Combustion: Applying Results of a Large 30 MWth Pilot Project. In: 9th International Conference on Greenhouse Gas Control Technologies; Washington, D.C., U.S.A.; 16-20 November 2008; Eds.: Washington, D.C., U.S.A., **2008**.

[10]   Total.Project Information Dossier. Lacq CO$_2$ capture and storage Pilot Project, **2007**.

[11]   Sturgeon, D. Oxyfuel Combustion Technology - Carbon Sequestration Leadership Forum. In: Official Opening of the OxyCoal^{TM} Clean Combustion Test Facility. Technical Seminar; 24 July **2009**.

[12]   Ochs, T.; Oryshchyn, D.; Woodside, R.; Summers, C.; Patrick, B.; Gross, D.; Schoenfield, M.; Weber, T.; O'Brien, D. Results of initial operation of the Jupiter Oxygen Corporation oxy- fuel 15 MWth burner test facility. Energy Procedia, **2009**, 1, 511-518.

[13]   Barbucci, P.CO$_2$ Capture and Storage- An utility view, In: 3rd International Symposium on Capture and geological storage of CO$_2$; Paris, France, 5-6 November 2009; ed.: Paris, France; **2009**.

[14]   Wall, T.; Liu, Y.; Spero, C.; Elliott, L.; Khare, S.; Rathnam, R.; Zeenathal, F.; Moghtaderi, B.; Buhre, B.; Sheng, C.; Gupta, R.; Yamada, T.; Makino, K.; Yu, J. An overview on oxyfuel coal combustion--State of the art research and technology development. Chem Eng ResDes, **2009**, 87, 1003-1016.

[15]   Scheffknecht, G.Research Perspective- Review of the Current Understanding, Identifying Research Gaps, In: 1st Oxyfuel Combustion Conference; Cottbus, Germany; 8-11 September 2009; ed.: Cottbus, Germany, **2009**.

[16] Spero, C.; Montagner, F. Oxy-combustion Technology in the World Scene, In: Coal 21 Conference; NSW, Australia; 18-19 September 2007; ed.: NSW, Australia; **2007**.

[17] Nsakala, N.; Liljedahl, G.N.; Turek, D.G. Greenhouse gas emissions control by oxygen firing in Circulating Fluidized Bed Boilers. Phase II-Pilot scale testing and updated performance and economics for oxygen fired CFB with CO_2 capture. Final Technical Progress Report, ALSTOM Power Inc. **2004**.

[18] Saastamoinen, J.; Tourunen, A.; Pikkarainen, T.; Häsä, H.; Miettinen, J.; Hyppänen; Myöhänen, K. Fluidized bed combustion in high concentrations of O_2 and CO_2, In: 19th FBC Conference; Vienna, Austria; 21-24 May 2006; ed.: Vienna, Austria, **2006**.

[19] Seddighi, S.; Pallarès, D.; Johnsson, F. One-dimensional modeling of oxy-fuel fluidized bed combustion for CO_2 capture, In: Fluidization XIII: New Paradigm in Fluidization Engineering; Gyeong-ju, Korea;2010 May 16-21; ed.: Gyeong-ju, Korea, **2010**.

[20] Jia, L.; Tan, Y.; Wang, C.;Anthony, E.J. Experimental Study of Oxy-Fuel Combustion and Sulfur Capture in a Mini-CFBC. Energ Fuel, **2007**, 21, 3160-3164.

[21] Pikkarainen, T.Small scale fluidized bed experiments under oxygen combustion conditions, In: 2007 International Conference on Coal Science and Technology; Nottingham, UK; 28-31August 2007; ed.: Nottingham, UK; **2007**.

[22] Czakiert, T.; Sztekler, K.; Karski, S.; Markiewicz, D.; Nowak, W. Oxy-fuel circulating fluidized bed combustion in a small pilot-scale test rig. Fuel Process Technol, **2010**, 91, 1617-1623.

[23] Krzywanski, J.; Czakiert, T.; Muskala, W.; Sekret, R.; Nowak, W. Modeling of solid fuel combustion in oxygen-enriched atmosphere in circulating fluidized bed boiler: Part 2. Numerical simulations of heat transfer and gaseous pollutant emissions associated with coal combustion in O_2/CO_2 and O_2/N_2 atmospheres enriched with oxygen under circulating fluidized bed conditions. Fuel Proces Technol, **2010**, 91, 364-368.

[24] de Diego, L.F.; de las Obras-Loscertales, M.; García-Labiano, F.; Rufas, A.; Abad, A.; Gayán, P.; Adánez, J. Characterization of a limestone in a batch fluidized bed reactor for sulfur retention under oxy-fuel operating conditions. International Journal of Greenhouse Gas Control, **2011**, 5, 1190-1198.

[25] Eddings, E.G.; Okerlund, R.; Bool, L.E. Pilot-scale evaluation of oxycoal firing in circulating fluidized bed and pulverized coal-fired test facilities, In: 1st Oxyfuel Combustion Conference; Cottbus, Germany; 8-11September 2009; ed.: Cottbus, Germany; **2009**.

[26] Varonen, M.4 MWth Oxy-CFB Test Runs, In: 63rd IEA FBC meeting; Ponferrada, Spain; 29-30 November 2011; ed.: Ponferrada, Spain, **2011**.

[27] Seddigh, S.; Pallarès, D.; Johnsson, F.; Varonen, M.; Hyytiäinen, I.; Ylä-Outinen, V.; Palonen, M. Assessment of Oxyfuel Circulating Fluidized Bed Boilers – Modeling and Experiments in a 5 MW Pilot Plant, In: 2nd Oxyfuel Combustion Conference; Queensland, Australia; 12-16 September 2011; ed.: Queensland, Australia; **2011**.

[28] Tondl, G.; Penthor, S.; Wöß, D.; Pröll, T.; Höltl, W.; Rohovec, J.; Hofbauer, H. From Oxygen enrichement to Oxyfuel combustion, In: 63rd IEA FBC meeting; Ponferrada, Spain; 29-30 November 2011; ed.: Ponferrada, Spain, **2011**.

[29] Scala, F.; Chirone, R.; Salatino, P. Recent Research on Fluidized bed Oxy-fuel combustion at Naples, In: 63rd IEA FBC meeting, Ponferrada, Spain, 29-30 November 2011; ed.: Ponferrada, Spain, **2011**.

[30]   Duan, L.; Zhao, C.; Zhou, W.; Qu, C.; Chen, X. O$_2$/CO$_2$ coal combustion characteristics in a 50 kWth circulating fluidized bed. International Journal of Greenhouse Gas Control, **2011**, 5, 770-776.

[31]   Guedea, I.; Díez, L.I.; Pallarés, J.; Romeo, L.M. Influence of O$_2$/CO$_2$ mixtures on the fluid-dynamics of an oxy-fired fluidized bed reactor. Chem Eng, **2011**, 178, 129-137.

[32]   Romeo, L.M.; Díez, L.I.; Guedea, I.; Bolea, I.; Lupiáñez, C.; González, A.; Pallarés, J.; Teruel, E. Design and operation assessment of an oxyfuel fluidized bed combustor. Exp Therm Fluid Sci, **2010**, 35, 477-484.

[33]   Lupiáñez, C.; Guedea, I.; Bolea, I.; Pallarés, J.; Díez, L.I.; Romeo, L.M. Oxy-firing of high sulphur coal in CIRCE fluidized bed pilot plant, In: 5th International Conference on Clean Coal Technologies; Zaragoza, Spain; 8-11 May 2011;ed.: Zaragoza, Spain; **2011**.

[34]   Toftegaard, M.B.;Brix, J.; Jensen, P.A.; Glarborg, P.; Jensen, A.D. Oxy-fuel combustion of solids fuels. Prog Energ Combust, **2010**, 36, 581-625.

[35]   Pohlmann, J.G.; Osorio, E.; Vilela, A.C.F.; Borrego, A.G. Reactivity to CO$_2$ of chars prepared in O$_2$/N$_2$ and O$_2$/CO$_2$ mixtures for pulverized coal injection (PCI) in blast furnace in relation to char petrographic characteristics. International Journal of Coal Geology, **2010**, 84, 293-300

[36]   Rathnam, R.K.; Elliott, L.K.; Wall, T.F.; Liu, Y.; Moghtaderi, B. Differences in reactivity of pulverised coal in air (O$_2$/N$_2$) and oxy-fuel (O$_2$/CO$_2$) conditions. Fuel Process Technol, **2009**, 90, 797-802.

[37]   Brix, J.; Jensen, P.A.; Jensen, A.D. Coal devolatilization and char conversion under suspension fired conditions in O$_2$/N$_2$ and O$_2$/CO$_2$ atmospheres. Fuel, **2010**, 89, 3373-3380.

[38]   Shaddix, C.R.; Molina, A. Particle imaging of ignition and devolatilization of pulverized coal during oxy-fuel combustion. Proceedings of the Combustion Institute, **2009**, 32, pp. 2091-2098.

[39]   Krzywanski, J.; Czakiert, T.; Muskala, W.; Nowak, W. Modelling of CO$_2$, CO, SO$_2$, O$_2$ and NO$_x$ emissions from the oxy-fuel combustion in a circulating fluidized bed. Fuel Process Technol, **2011**, 92, 590-598.

[40]   Molina, A.; Shaddix, C.R. Ignition and devolatilization of pulverized bituminous coal particles during oxygen/carbon dioxide coal combustion. Proceedings of the Combustion Institute, **2007**, 31, pp. 1905-1912.

[41]   Bejarano, P.A.; Levendis, Y.A. Single-coal-particle combustion in O$_2$/N$_2$ and O$_2$/CO$_2$ environments. Combust Flame, **2008**, 153, 270-287.

[42]   Li, Q.; Zhao, C.; Chen, X.; Wu, W.; Lin, B. Properties of char particles obtained under O$_2$/N$_2$ and O$_2$/CO$_2$ combustion environments. Chem Eng Process: Process Intensification, **2010**, 49, 449-459.

[43]   Jia, L.; Tan, Y.; McCalden, D.; Wu, Y.; He, I.; Symonds, R.; Anthony, E.J. Commissioning of a 0.8 MWth CFBC for oxy-fuel combustion, In: CFB-10; Oregon, USA; 1-5 May 2011; ed.: Oregon, USA; **2011**.

[44]   Krzywanski, J.; Czakiert, T.; Muskala, W.; Nowak, W. Modelling of CO$_2$, CO, SO$_2$, O$_2$ and NO$_x$ emissions from the oxy-fuel combustion in a circulating fluidized bed. Fuel Process Technol, **2011**, 92, 590-596.

[45]   Czakiert, T.; Bis, Z.; Muskala, W.; Nowak, W. Fuel conversion from oxy-fuel combustion in a circulating fluidized bed. Fuel Process Technol, **2006**, 87, 531-538.

[46] Jia, L.; Tan, Y.; Anthony, E.J. Emissions of SO₂ and NOₓ during Oxy–Fuel CFB Combustion Tests in a Mini-Circulating Fluidized Bed Combustion Reactor. Energ Fuel, **2009**, 24, 910-915.

[47] Gungor, A.; Eskin, N. Two-dimensional coal combustion modeling of CFB. Int J Therm Sci, **2008**, 47, 157-174.

[48] Oka, N.S.Fluidized Bed Combustion. New York: Marcel Dekker,Inc, **2004**.

[49] Oka, S.N., Fluidized Bed Combustion: Marcel Dekker Inc., **2003**.

[50] Kunii, D.; Levenspiel, O. Fluidization Engineering, **1991**.

[51] Basu, P., Combustion and gasification in fluidized beds, Eds.: Taylor & Francis, **2006**.

[52] Liu, H.; Katagiri, S.; Kaneko, U.; Okazaki, K. Sulfation behavior of limestone under high CO₂ concentration in O₂/CO₂ coal combustion. Fuel **2000**, 79, 945-953.

[53] Scala, F.; Salatino, P. Flue gas desulfurization under simulated oxyfiring fluidized bed combustion conditions: The influence of limestone attrition and fragmentation. Chem Eng Sci, **2010**, 65, 556-561.

[54] Anthony, E.J.; Granatstein, D.J. Sulfation phenomena in fluidized bed combustion systems. Prog Energ Combust **2001**, 27, 215-236.

[55] Fuertes, A.B.; Velasco, G.; Fuente, E.; Alvarez, T. Study of direct sulfation of limestone at high CO₂ partial pressures. Fuel Process Technol, 1994, 38,

[56] Fuertes, A.B.; Velasco, G.; Fuente, E.; Parra, J.B.; Alvarez, T. Sulphur retention by limestone particles under PFBC conditions. Fuel Process Technol, **1993**, 36,

[57] Jia, L.; Tan, Y.; Anthony, E.J. Emissions of SO₂ and NOₓ during Oxy–Fuel CFB Combustion Tests in a Mini-Circulating Fluidized Bed Combustion Reactor. Energ Fuel, **2010**, 24, 910-915.

[58] Jia, L.; Tan, Y.; McCalden, D.; Wu, Y.; He, I.; Symonds, R. and Anthony, E.J. Commissioning of a 0.8 MWth CFBC for oxy-fuel combustion. International Journal of Greenhouse Gas Control, **2012**, 7, 240-243.

[59] Duan, L.; Zhao, C.; Zhou, W.; Liang, C.; Chen, X. Sulfur evolution from coal combustion in O₂/CO₂ mixture. J Anal Appl Pyrol, **2009**, 86, 269-273.

[60] Duan, L.; Zhao, C.; Zhou, W.; Qu, C.; Chen, X. O₂/CO₂ coal combustion characteristics in a 50kWth circulating fluidized bed. International Journal of Greenhouse Gas Control, **2011**, 5, 770-776.

[61] Lupiáñez, C.; Scala, F.; Salatino, P.; Romeo, L.M.; Díez, L.I. Primary fragmentation of limestone under oxy-firing conditions in a bubbling fluidized bed. Fuel Process Technol, **2011**, 92, 1449-1456.

[62] Scala, F.; Salatino, P. Limestone fragmentation and attrition during fluidized bed oxyfiring. Fuel, **2010**, 89, 827-832.

[63] de Diego, L.F.; de las Obras-Loscertales, M.; Rufas, A.; García-Labiano, F.; Gayán, P.; Abad, A.; Adánez, J. Pollutant emissions in a bubbling fluidized bed combustor working in oxy-fuel operating conditions: Effect of flue gas recirculation. Appl Energ, **2013**, 102, 860-867

[64] de Diego, L.F.; Rufas, A.; García-Labiano, F.; de las Obras-Loscertales, M.; Abad, A.; Gayán, P.; Adánez, J. Optimum temperature for sulphur retention in fluidised beds working under oxy-fuel combustion conditions. Fuel, **2012**,

[65] García-Labiano, F.; Rufas, A.; de Diego, L.F.;de las Obras-Loscertales, M.; Gayán, P.; Abad, A.; Adánez, J. Calcium-based sorbents behaviour during sulphation at oxy-fuel fluidised bed combustion conditions. Fuel, **2011**, 90, 3100-3108.

[66]   Lupiáñez, C.; Guedea, I.; Bolea, I.; Díez, L.I.; Romeo, L.M. Experimental study of SO_2 and NO_x emissions in fluidized bed oxy-fuel combustion. Fuel Process Technol, **2013**, 106, 587-594.

[67]   Leckner, B. Fluidized bed combustion: Mixing and pollutant limitation. Prog Energ Combust, **1997**, 24, 31-61.

[68]   Duan, L.; Zhao, C.; Zhou, W.; Qu, C.; Chen, X. Effects of operation parameters on NO emission in an oxy-fired CFB combustor. Fuel Process Technol, **2011**, 92, 379-384.

[69]   Duan, L.; Zhao, C.; Ren, Q.; Wu, Z.; Chen, X. NO_x precursors evolution during coal heating process in CO_2 atmosphere. Fuel, **2011**, 90, 1668-1673.

[70]   Krzywanski, J.; Czakiert, T.; Muskala, W.; Sekret, R.; Nowak, W. Modeling of solid fuels combustion in oxygen-enriched atmosphere in circulating fluidized bed boiler: Part 1. The mathematical model of fuel combustion in oxygen-enriched CFB environment. Fuel Process Technol, **2010**, 91, 290-295.

[71]   Wilke, C.R. A Viscosity Equation for Gas Mixtures. The Journal of Chemical Physics, **1950**, 18,

[72]   Baskakov, A.P.; Leckner, B. Radiative heat transfer in circulating fluidized bed furnaces. Powder Technol, **1997**, 90, 213-218.

[73]   Howard, J.R.Fluidized Bed Technology: principles and applications. Philadelphia, Eds.: Taylor & Francis, **1989**.

CHAPTER 2

Gas Separation Membranes Used in Post-Combustion Capture

Li Zhao[*] and Ludger Blum

Institute of Energy and Climate Research, Fuel Cells (IEK-3), Leo-Brandt-Straße, Forschungs Zentrum Jülich, GmbH, D-52425 Jülich, Germany

Abstract: During the utilization of CO_2/N_2 gas separation membranes for post-combustion capture, the most important problem is how to create the driving force efficiently because the feed flue gas has only ambient pressure and a relatively low CO_2 content. Multi-stage systems are necessary using feasible membranes in order to fulfill the separation target, required by the following pipeline transport, and limited by the storage capacity. The whole work was divided into two steps: energy consumption analysis and capture cost analysis.

This book chapter describes mass and energy balances for single-stage and multi-stage membrane systems used in coal-fired power plant. After the recirculation of flue gas and variation of the feed gas compressor and vacuum pump on the permeate side, two concepts were developed and optimized to achieve minimum energy consumption. In order to evaluate different membrane capture concepts, a comparison with chemical absorption process was carried out, considering different degrees of CO_2 separation. Furthermore, a cost model was developed to make further analysis of the optimized concept in view of the tradeoff balance between material and energy consumption. The correlation between the membrane parameters (selectivity, permeability) and capture cost was investigated.

Keywords: CCS, post-combustion, gas separation membrane, multi-stage, energy consumption, economic analysis.

1. INTRODUCTION

Global warming has been identified as one of the world's major environmental issues. The relation between anthropogenic emissions of CO_2 and increased atmospheric CO_2 levels, and the associated high global temperatures, has been

*Corresponding author Li Zhao: Institute of Energy and Climate Research, Fuel Cells (IEK-3), Leo-Brandt-Straße, Forschungs Zentrum Jülich, GmbH, D-52425 Jülich, Germany; Tel: 02461/614064; Fax: 02461/616695; E-mail: l.zhao@fz-juelich.de

well established and accepted throughout the world. Carbon dioxide capture and storage (CCS) technologies constitute a promising option that can drastically reduce CO_2 emissions. CCS is a process causing the separation of CO_2 from industrial and energy-related sources, and transport to a storage location and long-term isolation from the atmosphere. CO_2 can be captured by a variety of methods, which can be classified as post-combustion, pre-combustion and oxy-combustion [1].

Compared with other scenarios of CO_2 capture and storage (CCS), *i.e.* pre-combustion and oxyfuel, an eminent feature of post-combustion capture is that it can be applied to existing plants, although it does require a corresponding technical modification. Chemical absorption with amines, such as monoethanolamine (MEA), is the process of choice at the present time [2, 3]. Although the chemical absorption method occupies a leading position in R&D on post-combustion with CCS, it has several innate weaknesses such as: a) degradation of the solvent, leading to high material costs, high disposal costs and additional environmental pollution; b) high energy consumption - almost 50% of the low-pressure steam from the intermediate/low pressure (IP/LP) steam turbine would be employed for regeneration of the solvent used to release CO_2; c) auxiliary retrofitting measures or costs in power plants - on the one hand, in order to extend the solvent lifetime, the high demands caused by the SO_x (< 10 ppm) and NO_2 (< 20 ppm) content in the flue gas lead to extraordinarily high expenditure for the pre-capture process; on the other hand, for the existing power plant calculations should include the fact that the modification of the IP/LP steam turbine is unavoidable because of the extraction of steam for CO_2 regeneration [2, 4-6].

As a technology competing with MEA absorption, the CO_2/N_2 gas separation membrane process for post-combustion capture is attracting more and more attention around the world [7-10]. In comparison with the above-mentioned weaknesses of chemical absorption, CO_2 gas separation membranes possess the following advantages: a) Membranes have potential for less environmental impact; b) Membrane modules can be used as add-on equipment in power plants; c) Membrane module is compact and flexible. These are eminently important properties of the gas separation membrane process distinguishing it from the other post-combustion capture technologies.

In this book chapter, different membrane processes have been investigated on the basis of energetic analysis. A comparison between the membrane and MEA absorption capture process has been carried out, considering different degrees of CO_2 separation. Furthermore, a cost model was developed to make a further analysis of the optimized concept in view of the tradeoff balance between material and energy consumption. The correlation between the membrane parameters (selectivity, permeability) and capture cost was investigated.

2. CO$_2$ SELECTIVE MEMBRANES

The current membranes applicable for gas separation in post-combustion processes can be mainly divided into inorganic ceramic membranes and organic polymeric membranes. So-called hybrid membranes are composed of inorganic molecular sieves and polymers [11] or, *vice versa*, microporous polymer layers supported on porous ceramic substrates [12, 13].

For inorganic ceramic membranes for CO_2 separation from flue gas, the US Department of Energy (DOE) requires a CO_2 permeance $> 3 \times 10^{-7}$ mol m^{-2}s^{-1}Pa^{-1} (2.42 m^{3}m^{-2}h^{-1}bar^{-1}) and a CO_2/N_2 selectivity > 100 [14]. This is supported by NEDO (New Energy and Industrial Technology Development Organization) in Japan. The primary goals of the NEDO group are to develop an inorganic ceramic membrane with the following properties: CO_2/N_2 selectivity of 100 and permeance of CO_2 at 3.4×10^{-7} mol m^{-2}s^{-1}Pa^{-1} (2.74 m^{3}m^{-2}h^{-1}bar^{-1}) at 350 °C, also with a durability of 500 hours at 350 °C [15]. The CO_2 separation membrane could be either a single or composite structure made of *e.g.* silica, zeolite, zirconia, titania or alumina [15-18]. The state-of-the-art ceramic membrane has quite good CO_2 permeance larger than 2×10^{-7} mol m^{-2}s^{-1}Pa^{-1} (1.61 m^{3}m^{-2}h^{-1}bar^{-1}) [18], but the selectivity is not above 10 [14, 18].

With regard to polymeric membranes, Powell and Qiao [10] made a detailed review of different polymer structures developed for flue gas separation. The original discussion of the "Robeson plot" has become one of the most highly cited papers in the gas membrane separation literature [19]. A comparison to calculation results from Freeman [20] and a non-exhaustive collection of experimental selectivity/ permeability data for CO_2/N_2 separation cited from [21] are shown in Fig. **1**.

The current potential polymer membranes can be categorized as following [10, 22, 23]:

- Polyimides: Polymides are very attractive materials for gas separation membranes because of their good gas separation and physical properties, such as high thermal stability, chemical resistance, mechanical strength, and low dielectric constant. Some polyimides, particularly those incorporating the group 6FDA possess very high carbon dioxide combined with high selectivity. An extensive review of the gas separation properties of polyimides was published by Langsam in 1996 [24]. However, problems associated with the swelling and plasticization of polyimides can limit its applications in the gas separation area.

- Facilitated transport: These membranes comprise a carrier (typically, metal ions) with a special affinity towards a target gas molecule and this interaction controls the rate of transport. Facilitated transport membranes can be divided into two general categories: fixed carrier (chained carrier) and mobile carrier (immobilized liquid) membranes. The relevant representative membranes are polyamidoamine (PAMAM) dendrimers [25] and polyvinyl amine (PVAm)/polyvinyl alcohol (PVA) [26], respectively. Facilitated transport membranes have received a lot of attention in gas separations because they offer high selectivity and large fluxes.

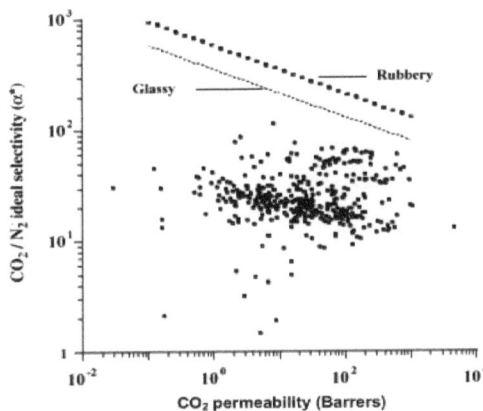

Figure 1: Robeson plot, trade-off curves for glassy (dotted line) and rubbery (bold dotted line) polymers are taken as the upper bound for polymer membrane separation ability; points represent a collection of experimental selectivity/permeability data for CO_2/N_2 mixtures [21].

- Mixed matrix: Their microstructure consists of an inorganic material in the form of micro- or nano-particles (discrete phase) incorporated into a polymeric matrix (continuous phase). Koros *et al.,* [27, 28] proposed some criteria for material selection and preparation protocols in order to match the necessary transport characteristics of materials to prepare high performance mixed matrix membranes for CO_2 capture. However, their performance suffers from defects caused by poor contact at the inorganic/polymer interface.

- Carbon molecular sieves: They are obtained through the pyrolysis (at high temperature in an inert atmosphere) of polymeric (mostly Polyimide) precursors already processed in the form of membranes. Carbon membranes [29-31] combine improved gas transport properties for light gases (gases of molecular size smaller than 4.0-4.5 Å) with thermal and chemical stability. The major disadvantages that hinder their commercialization are their brittleness and the cost which is 1 to 3 orders of magnitude greater per unit area than polymeric membranes.

- Poly-ether-oxide (PEO): PEO membranes are considered as attractive materials for CO_2 separation owing to the fact that the polymer chain has so strong affinity to CO_2 molecule for the presence of polar ether oxygen. There have been numerous efforts to design polymers containing poly (ethylene oxide) (PEO) for CO_2/N_2 and CO_2/H_2 separations in part because ethylene oxide units have a high concentration of ether oxygen and are relatively easy to fabricate. However, PEO is also subject to a strong tendency to crystallize, which is deleterious for gas permeability. Therefore, various techniques have been applied to reduce crystallinity in PEO, *e.g.* using low molecular weight liquid PEO or polyethylene glycol (PEG). The polymer membranes developed by GKSS [32-36], Germany are used here.

3. ENERGETIC ANALYSIS FOR MEMBRANE PROCESSES

It is well known that the driving force of gas separation membranes for CO_2 permeation is the partial pressure difference between the feed and the permeate side. Owing to the limitations of the operating conditions of post-combustion capture, a certain CO_2 partial pressure difference must be created in order to obtain an adequate driving force. The corresponding measures can be taken: on the feed side, 1. increasing the pressure (using a compressor); 2. increasing the CO_2 concentration (recirculating an enriched CO_2 stream); on the permeate side, 3. decreasing the pressure (using a vacuum pump); 4. decreasing the CO_2 concentration (using sweep gas).

The application of a compressor and vacuum pump leads to electrical energy consumption, which causes the energy penalty of existing power plants. One aspect to be strengthened here is that by coupling an expander on the retentate side part of the energy used by the compressor on the feed side can be recovered, whereas this is impossible if a vacuum pump is used on the permeate side. Using sweep gas on the permeate side could be the most energy-saving concept, but it is not easy to find a feasible source and mass in the post-combustion process. Using sweep gas on the permeate side (four-end membrane: feed, retentate + sweep gas, permeate) could be an energy-saving concept and has already been investigated by some researchers [37, 38]. However, ways of finding a feasible source for a sweep gas in the post-combustion process and how to separate the sweep gas and permeate in order to obtain the desired permeate product purity should be investigated in more depth. This paper focuses on compression machines (three-end membrane: feed, retentate + permeate).

3.1. Reference Power Plant and Simulation Method

In the present work, a reference power plant termed the Reference Power Plant North Rhine-Westphalia (RKW-NRW) [39] was chosen for energy analyses. The multi-stage polymer membranes should be installed after the SCR-DeNOx, dust removal (E-filter) and desulphurization (FGD) processes and before emissions pass through the cooling tower, analogous to amine stripping processes [2]. The position of CO_2 membrane capture in a power plant is schematically illustrated in Fig. **2**. At this position, the flue gas has 1 bar pressure and temperature of 50~70 °C.

The hard coal grade "Klein Kopje" was used to simulate the flow rate and the components of the flue gas for the multi-stage membrane calculation. The elemental analysis data of Klein Kopje coal are: C 65.5%, H 3.5%, O 7.4%,

N 1.5%, S 0.6%, ash 14.2%, moisture 7.3%; and the heat value is 25 MJ/kg. The coefficient of air excess (air-to-fuel ratio) was assumed to be 1.15. The basic data of RKW-NRW and the simulation results of the flue gas are listed in Table **1**. The residue of the pollutant in the flue gas consists of approximately 50 vppm SO_2 and approximately 200 ppm NO_2.

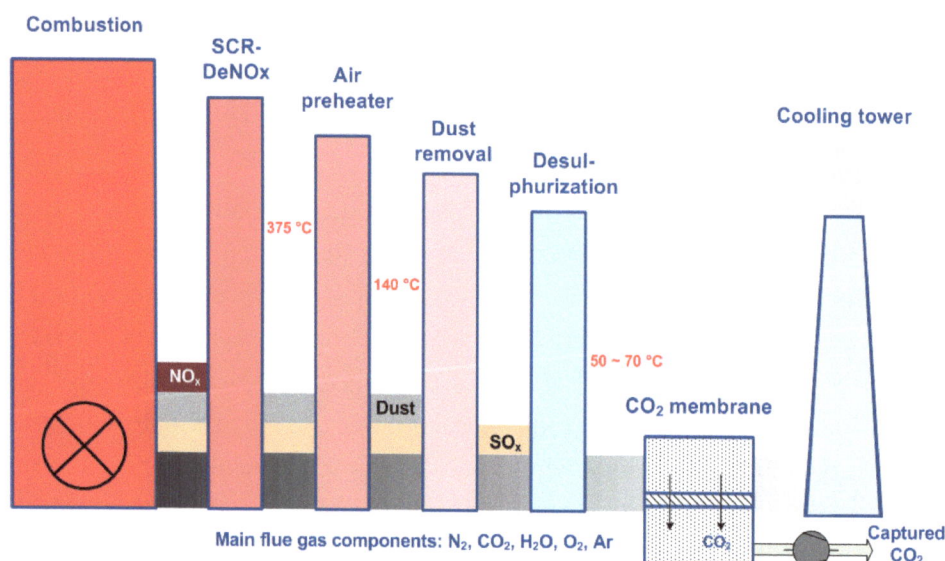

Figure 2: Schematic diagram of CO_2 membrane position in a post-combustion flue gas line.

The PRO/II (Simulation Science Inc.) software was used for the simulation. There are different thermodynamic models for the energy balance calculation available in PRO/II; for the case described here the Soave-Redlich-Kwong equation of state was chosen. The adiabatic efficiency of the compressors, expanders and vacuum pumps is assumed to be 85%. A binary flue gas system - 14 mol% CO_2 and 86 mol% N_2 were simulated.

3.2. Single-Stage Membrane System

A single-stage membrane process is shown in Fig. **3**. All relevant parameters are listed here. Operating conditions (pressure, temperature, CO_2 concentration and flow rate of the feed gas) act on a certain membrane (selectivity, permeability and area), then the performance of the membrane (CO_2 purity and degree of CO_2 separation) can be predicted.

Table 1: RKW-NRW power plant basic data [39] and simulation results of the flue gas conditions after removal of the pollutants using Klein Kopje hard coal

Power Plant RKW-NRW	
Output gross	600 MW
Output net	555 MW
Net efficiency	45.9%
Steam parameters	285 bar/600 °C/620 °C
Operation time	6000 h/year
Fuel input	1.0 Mt/year[*]
Investment costs	517.1 million euro
O & M costs	7.8 million euro/year
Fuel costs	41 euro/t
Electricity price	3.37 cent/kWh
Flue Gas Conditions After Removal of the Pollutants	
Pressure	1.05 bar
Temperature	50 °C
Flow rate	1.6 million m^3/h[*]
Main Components	
CO_2	13.5 mol%[*]
N_2	70.1 mol%[*]
O_2	3.7 mol%[*]
H_2O	11.9 mol%[*]
Ar	0.8 mol%[*]

* simulated by PRO/II

Considering the correlation among these parameters, CO_2 purity, degree of CO_2 separation, membrane area and specific energy, two diagrams were developed for a single-stage membrane using a vacuum pump on the permeate side, shown in Figs. **4a** and **4b**.

The lines of filled squares indicate the correlation between CO_2 purity and degree of CO_2 separation with increasing membrane area. The lines of filled dots show how the specific energy increases with the enhancive degree of CO_2 separation. The vacuum level varies from 200 mbar to 30 mbar, as marked by different colors in the diagram (200 mbar: brown, 150 mbar: pink, 100 mbar: green, 70 mbar: blue, 50 mbar: red, 30 mbar: black). In order to better understand the following investigation results, some points concerning the required membrane area are marked in the diagram, *e.g.* if the permeate vacuum is 30 mbar, to reach the 50% degree of CO$_2$ separation 161 m^2 of membrane is needed for the 100 Nm^{3}h^{-1} feed gas flow rate. It is obvious that a high degree of permeate vacuum contributes to a high CO$_2$ purity and to a high degree of separation, but leads to a higher energy consumption for the process.

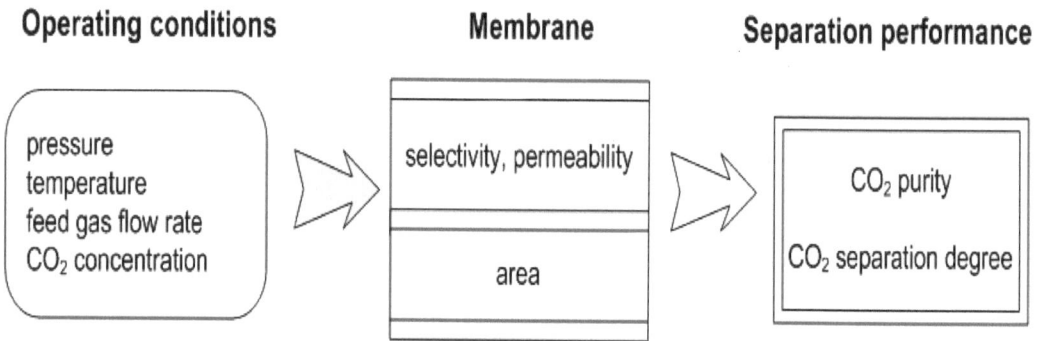

Figure 3: All relevant parameters of a single-stage membrane process.

The advantage of these diagrams is that the relevant specific energy and CO$_2$ purity for the same process can be obtained simultaneously, *e.g.* for the above-mentioned case, the CO$_2$ purity is 80 mol% and the specific energy is 106 kWh/t$_{\text{separated CO}_2}$. A detailed parametric study was carried out in the paper [40].

For capture processes, one important factor must be considered, *i.e.* CO$_2$ purity, required by the following pipeline transport [41] and storage capacity. Furthermore, degree of CO$_2$ separation is the other important parameter. In order to minimize CO$_2$ emission, CO$_2$ should be separated as much as possible, which leads to a relevant high energy consumption and high investment cost.

Furthermore, large vacuum pumps will probably have a suction pressure of 50 mbar in the future, the pressure drop within the module channels and connecting tubing should be considered additionally. Under these boundary conditions Table **2** shows the analysis results of single-stage membrane separation performance. Then it can be deduced that using single-stage membrane is impossible to fulfill the separation targets, required in the CCS framework. Developing multi-stage membrane systems should be a potential option.

a)

b)

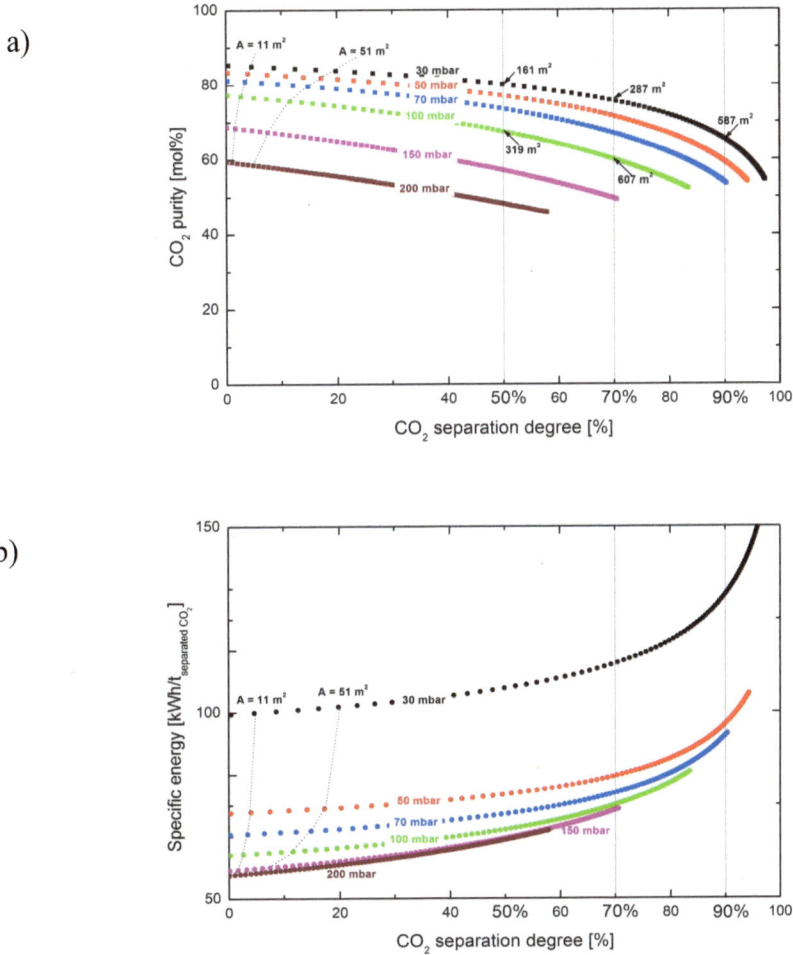

Figure 4: Correlation among the parameters for a single-stage membrane: a) Degree of CO$_2$ separation and CO$_2$ purity; b) Degree of CO$_2$ separation and specific energy; feed pressure 1.05 bar, feed gas flow rate 100 Nm^{3}h^{-1}, composition: 14 mol% CO$_2$, 86 mol% N$_2$, membrane selectivity α = 43, CO$_2$ permeance 0.5 Nm^{3}m^{-2}h^{-1}bar^{-1}, A = 1, 11, 21, ... 991 m^2.

3.3. Multi-Stage Membrane System

3.3.1. Cascade System

In our previous paper [42] a cascade membrane system has been developed and analyzed. Through varying the position of compressors and vacuum pumps, recycling retentate back to the feed stream, the cascade was optimized and shows energetic advantage with 50% and 70% degree of CO$_2$ separation in comparison with MEA absorption. Fig. **5** shows this cascade, using vacuum pump for the 1st membrane and compressor for the 2nd membrane, and recycling the retentate of the 2nd membrane to the feed stream. In the simulation, the polyactive membrane was used with CO$_2$/N$_2$ selectivity of 50 and CO$_2$ permeance of 3 Nm^{3}m^{-2}h^{-1}bar^{-1}.

Table 2: Separation performance of single-stage membrane under different vacuum levels

Permeate Vacuum [mbar]	CO$_2$ Separ. Degree [%]	Membrane	CO$_2$/N$_2$ Selectivity	CO$_2$ Purity [mol%]
30 not practicable	50	Polyactive	50	~80
		Theoretical	~200	95
100	50	Polyactive	50	~70
		Theoretical	3750	95
100	70	no solution		95
	90			95

Polyactive (GKSS): CO$_2$/N$_2$ selectivity = 50, CO$_2$ permeance: 3 Nm^{3}m^{-2}h^{-1}bar^{-1}; Theoretical membrane: CO$_2$ permeance: 3 Nm^{3}m^{-2}h^{-1}bar^{-1}.

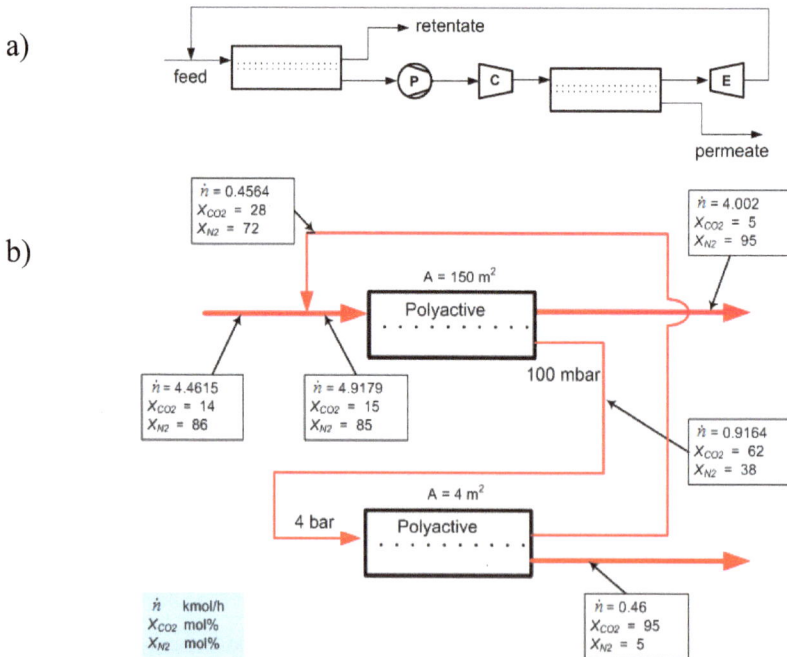

Figure 5: a) A cascade system, P-vacuum pump, C-compressor, E-expander; b) An example for this cascade (70% degree of CO$_2$ separation), feed gas flow rate 100 Nm^{3}h^{-1} (4.4615 kmolh^{-1}).

Fig. **6** shows the correlation between efficiency loss, specific energy and separation degree of cascade membrane, so as to MEA absorption. The green solid lines of filled square and dot stand for membrane cascade, and the red dashed lines for MEA absorption. An important feature of membrane cascade is that the specific energy consumption is not constant as that of MEA absorption, but changed with the degree of CO_2 separation. This provides an option that it can be used as a retrofit in existing power plants.

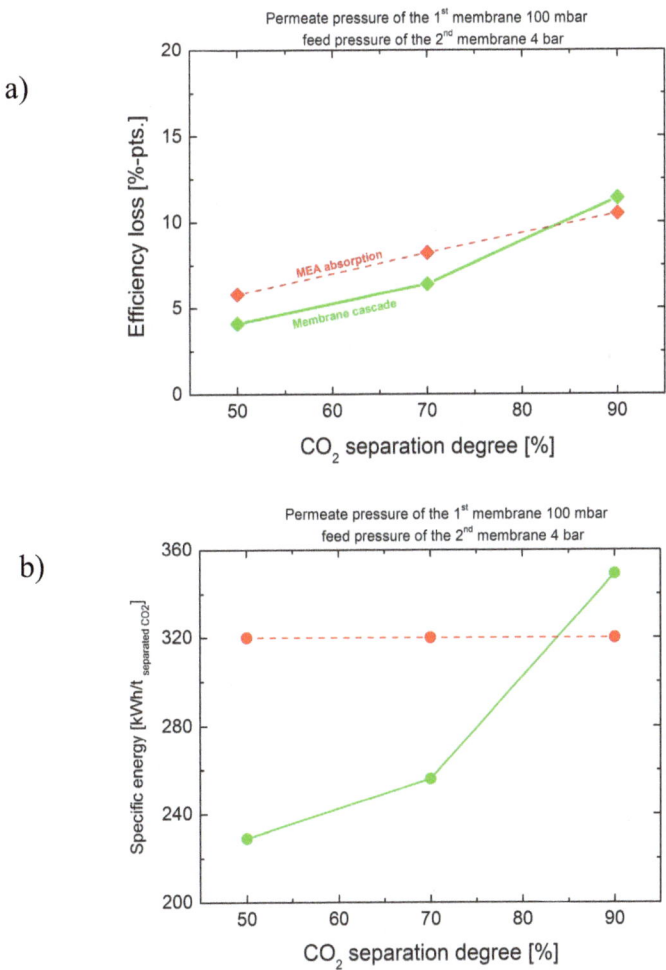

Figure 6: Correlation between a) Efficiency loss and separation degree, b) Specific energy and separation degree of the cascade membrane concept with MEA absorption in 50%, 70% and 90% degree of CO_2 separation, vacuum pump pressure of the 1st membrane 100 mbar, feed pressure of the 2nd membrane 4 bar, separated CO_2 compressed to 110 bar, 30 °C.

Furthermore, one factor cannot be neglected is the membrane area. Fig. **7** shows the increase of membrane area with the degree of CO_2 separation. The green line of filled square stands for the total membrane area, and the blue lines of hollow square stand for the area of the 1st membrane, and the blue line of hollow circle stands for the area of the 2nd membrane. It is obvious that the 1st membrane in this cascade plays a dominant role for the cascade separation and occupies most of the membrane area of the cascade.

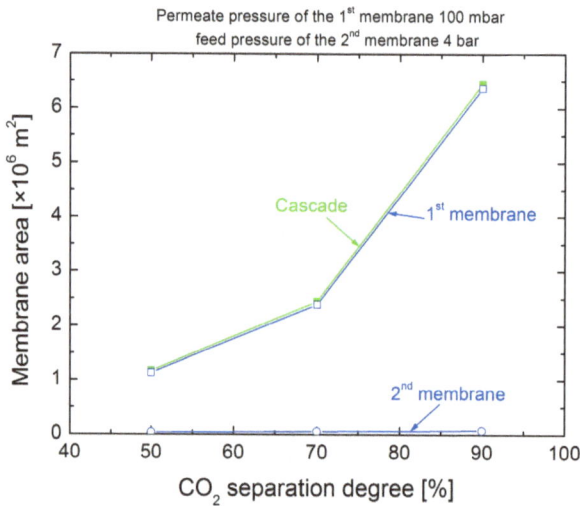

Figure 7: Correlation between membrane area and degree of CO_2 separation.

3.3.2. 2-Stage Cascade

In order to decrease large membrane area used for separation, a further development is using compressor for the feed gas. 2-stage cascades were developed for this purpose. Fig. **8** illustrates a variant in our previous paper [43].

Several aspects were considered for this variation: 1) Feasible large scale vacuum pump of 200 mbar level; 2) Using compressor for feed stream to decrease membrane area further; 3) 90% degree of CO_2 separation. Table **3** shows the simulation results under these boundary conditions. The specific energy of this variant is 356 kWh/ton$_{\text{separated CO}_2}$, and efficiency penalty is 12.3%-points. The state-of-art MEA absorption consumes 320 kWh/ton$_{\text{separated CO}_2}$, and efficiency penalty is 10.3%-points [44, 45].

a)

b)

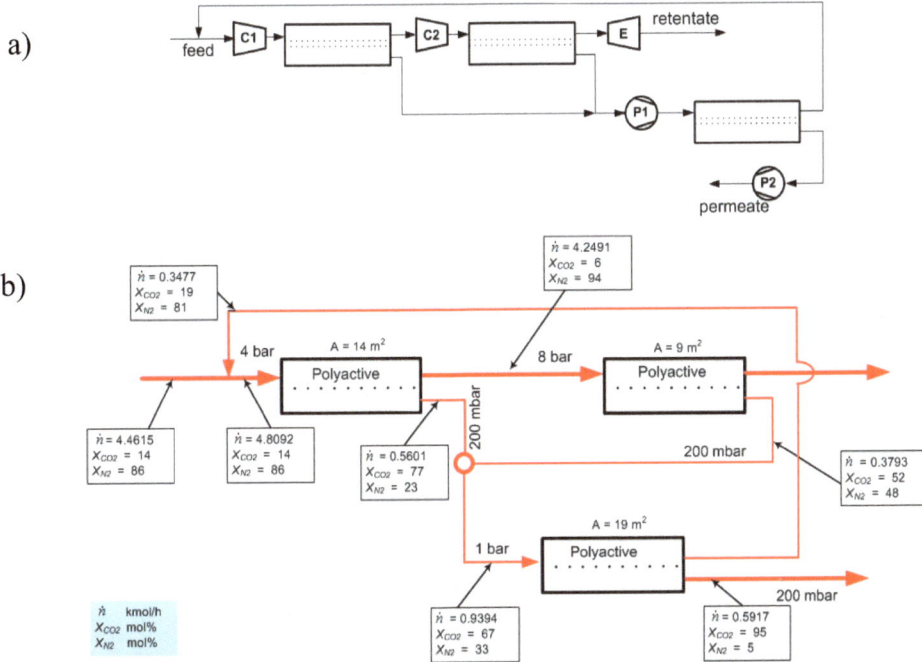

Figure 8: a) Using compressor for feed gas in 2-stage cascade, C1, C2: compressor, P, P1, P2: vacuum pump, E: expander; b) An example for this 2-stage cascade (90% degree of CO$_2$ separation), feed gas flow rate 100 Nm^{3}h^{-1} (4.4615 kmolh^{-1}).

Table 3: Simulation results of membrane area and energy consumption of 2-stage cascade, separated CO$_2$ compressed to 110 bar, 30 °C

Concept	Separation Degree [%]	CO$_2$ Purity [mol%]	Membrane Area [×10^6 m^2]	Energy Consumption [kWh/ton $_{separated\ CO2}$]
2-stage cascade	90	95	0.66	356
	Compressor pressure C1 = 4 bar, C2 = 8 bar, P1 = P2 = 200 mbar			

3.3.3. Using Sweep Gas

For multi-stage membrane concepts two developmental routes can be observed: three-end membrane (feed, retentate + permeate) and four-end membrane (feed, retentate + sweep gas and permeate).

Arshad Hussain *et al.,* [37] developed a cascade using water steam and permeate as sweep gas (Fig. **9**). The flue gas is compressed to the desired upstream pressure and cooled down prior being fed to the first membrane stage. The water vapors at very low pressure (25-125 mbar) are fed to the first membrane stage as a sweep. The flow rate of sweep (water vapor) is maintained at about 5% corresponding to the feed flow rate. As mentioned earlier, water vapor being used as sweep, not only enhance the membrane performance but also facilitate the efficient removal of permeate from the membrane stage, thus lowering the partial pressure of CO_2 on the permeate side. The mixture of permeate and water vapors exiting from the membrane is compressed to 2 bar and cooled at 3 °C before entering the separator/water knockout vessel, where CO_2 enriched gas stream is separated from water, heated to 25 °C and fed to second membrane stage for further CO_2 separation. It is vital to remove water from the permeate stream to avoid any corrosion/damage in the process machines. The water separated from permeate is heated by passing through a series of heat exchangers to acquire the vaporization temperature and then expanded to respective downstream pressure for its re-use as sweep.

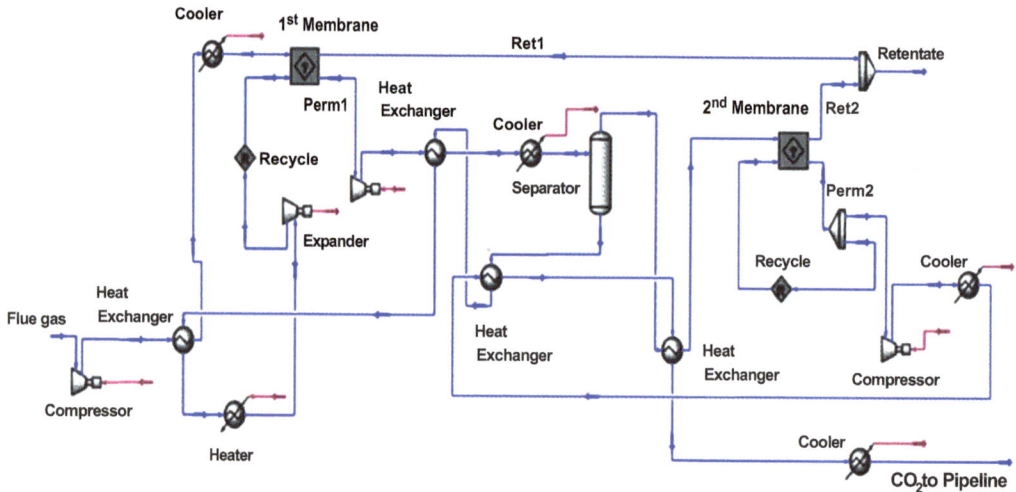

Figure 9: Using H_2O and permeate as sweep gas for membrane separation process [37].

Tim Merkel *et al.*, [38] developed a so called "two-step counter flow/sweep design" (Fig. **10**). In this scheme, a vacuum pump is used on the permeate side of the first membrane step (unit I). As discussed above, because the volume of the permeate gas (stream②) passing through the vacuum pump is only a fraction of the volume of the flue gas (stream①), the power used by the vacuum pump is much smaller than the power consumed by compressing the feed gas. This cross-flow membrane unit only removes a portion of the CO_2 in flue gas in a single pass, in order to reduce the membrane area and energy required in this step. The residue gas leaving the cross-flow membrane step (stream③) still contains about 7% CO_2. This gas passes on one side of a second membrane (unit II) that has a counter-flow/sweep configuration. All or a portion of the feed air to the boiler (stream④) passes on the permeate side of this membrane as a sweep stream. Because of the low CO_2 concentration in the air sweep, some CO_2 permeates through the membrane and is recycled with the feed air to the boiler (stream⑤). The treated flue gas (stream⑥) leaving the counter-flow membrane unit contains 1.8% CO_2 and is vented. So 90% CO_2 removal is achieved.

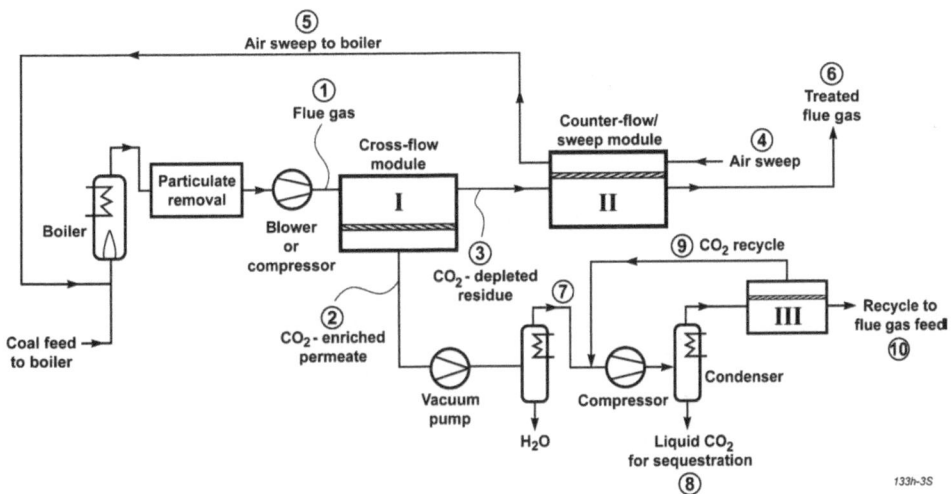

Figure 10: Using air as sweep gas for membrane separation process [38].

Comparing the three-end and four-end membrane concepts, some conclusions can be drawn:

- Using sweep gas on the permeate side could be an energy-saving concept, but it consumes still energy to separate the captured CO_2 and sweep gas. Furthermore, using sweep gas means that the membrane process must be integrated with some power plant components, *e.g.* the permeate with sweep gas of the 1st step membrane in Fig. **9** must be induced to the boiler; so the above mentioned "add-on" advantage of membrane capture process will be lost.

- 90% degree of CO_2 separation is a big challenge for any concept, which leads to 12~14% points efficiency loss;

- An important feature of three-end membrane capture process is, that for 50% and 70% degree of CO_2 separation the energy consumption is lower than that of MEA absorption capture. This shows an application potential for membrane capture as a retrofit for existing power plants.

4. CAPTURE COST ANALYSIS

4.1. Capture Cost

Applying a gas separation membrane system for post-combustion, the following cost factors should be considered: a) capital cost (including membrane, frame, compression equipment and heat exchanger); b) O&M cost; and c) energy consumption cost. An investigation of the literature [46-48] shows that the capture cost for MEA absorption lies in the range of 30~50 euro/t$_{\text{separated CO2}}$.

Referring to work by a Dutch group [47, 49], in the present work a similar simulation method was used to calculate the capture cost using the 600 MW NRW reference power plant. Table **4** lists 12 equations applied to determine the total capture cost and CO_2 specific separation cost; the relative cost and process parameters are shown in Table **5**.

The depreciation time for the components of compressor, expander, vacuum pump, heat exchanger and membrane module is 25 years, and the lifetime of the membrane is 5 years; the O&M cost for the components of compressor, expander, vacuum pump and heat exchanger is assumed to be 3.6% of their capital cost, and for the membrane and membrane frame the O&M cost is taken as 1% of their capital cost. Here the compressor cost was related to the flue gas of 2~8 bar and a vacuum pump of 200 mbar. It is assumed that the vacuum pump costs 4 times as much as the compressor ($K_{vp} = 4K_c$).

The electricity price here is 3.37 cent/kWh. One aspect to be emphasized is that this price is the current power cost. The capture cost calculated here shows the CO_2 separation expense using the existing infrastructure. The purpose of CCS is to reduce CO_2 emissions to the atmosphere. According to that view it is not the amount of carbon dioxide *captured* per unit of production (*e.g.* per kWh electricity) that is important, but it is the amount of carbon dioxide emission *avoided.* Due to CCS additional energy input is required per unit of output, if the power supply to the grid remains constant. As a result, the amount of CO$_2$ avoided is less than the amount of CO$_2$ captured, so that the cost per ton of CO$_2$ avoided will be higher than the cost per ton of CO$_2$ captured [51]. This indicates the direction of the further investigation.

Table 4: Equations applied to determine specific CO$_2$ separation cost [47, 49]

Estimated Investments I (Components)		
$I_m = A \cdot K_m \ \ldots$	(1)	Membrane cost
$I_{mf} = (A/2000)^{0.7} \cdot K_{mf} \ \ldots$	(2)	Permanent membrane frame cost
$I_c = K_c \cdot F_h \ \ldots$	(3)	Compressor cost
$I_{vp} = K_{vp} \cdot F_h \ \ldots$	(4)	Vacuum pump cost
$I_{ex} = P_{ex} \cdot K_{ex} \cdot F_h \ \ldots$	(5)	Expander cost
$I_{he} = C_{he} \ \ldots$	(6)	Heat exchangers and cooling facilities
Energy consumption of compression equipment *P*		
$P_{tot} = \sum P_c + \sum P_{vp} - \sum P_{ex} \ \ldots$	(7)	Total energy consumptio n
Annual costs C		
$C_{cap} = (\sum I_c + \sum I_{vp} + \sum I_{ex} + \sum I_{he} + I_{mf}) \cdot a + I_m \cdot a_m \ \ldots$	(8)	Capital cost
$C_{O\&M} = 0.036 \cdot (\sum I_c + \sum I_{vp} + \sum I_{ex} + \sum I_{he}) + 0.01 \cdot (I_m + I_{mf}) \ldots$	(9)	O&M cost
$C_{en} = t_{op} \cdot P_{tot} \cdot K_{el} \ \ldots$	(10)	Energy cost per year
$C_{tot} = C_{cap} + C_{en} + C_{O\&M} \ \ldots$	(11)	Total cost
Specific CO$_2$ separation cost $\dot{C}_{CO_2}$		
$\dot{C}_{CO_2} = C_{tot} / M_{CO_2,ann,separated} \ \ldots$	(12)	

Table 5: Assumptions for cost and process parameters [39, 47, 49, 50]

Parameter	Value	Parameter	Value
K_m	50	K_{mf}	0.25
K_c	30	K_{ex}	0.3
C_{he}	3.5	F_h	1.8
a	0.064	a_m	0.225
t_{op}	6000	K_{el}	3.37

Table 6: Capture costs for the cascade and 2-stage cascade systems

Cases		Cost		Unit	Value
Case 1: cascade Vacuum P 100 mbar Comp. C 4 bar	Case 1-1: **70%** separation degree	Specific CO$_2$ separation cost		euro/t$_{CO2}$/year	31
		Total cost		million euro/year	55.8
		Capital cost			35.3
			Membrane	million euro	122.2
			Membrane frame		36.2
			Compressor		59.4
			Vacuum pump		21.6
		O&M cost		million euro/year	4.7
		Energy cost			15.8
	Case 1-2: **90%** separation degree	Specific CO$_2$ separation cost		euro/t$_{CO2}$/year	49
		Total cost		million euro/year	114
		Capital cost			78.0
			Membrane	million euro	309.5
			Membrane frame		69.4
			Compressor		59.4
			Vacuum pump		21.6
		O&M cost		million euro/year	7.1
		Energy cost			27.1
Case 2: 2-stage cascade C1 4bar C2 8bar P1, P2 200 mbar	Case 2-1: 70% separation degree	Specific CO$_2$ separation cost		euro/t$_{CO2}$/year	44
		Total cost		Million euro/year	80.2
		Capital cost			43.3
			Membrane		25.2
			Membrane frame		12.0
			Compressor		113.4

Table 6: contd....

		Vacuum pump		432
		O&M cost	million euro/year	21.1
		Energy cost		15.8
	Case 2-2:	Specific CO_2 separation cost	euro/t $_{CO2}$/year	37
	90% separation degree	Total cost	Million euro/year	87.0
		Capital cost		45.0
		Membrane	million euro	32.3
		Membrane frame		14.3
		Compressor		113.4
		Vacuum pump		432
		O&M cost	million euro/year	21.2
		Energy cost		20.0

In this section, the analyses focus on the cascade (Fig. **5**) and 2-stage cascade (Fig. **8**) systems. On the basis of the capture cost simulation method described above, the CO_2 separation costs of these two systems are listed in Table **6**. Case 1-1 is the cascade with 70% degree of CO_2 separation and Case 1-2 with 90% degree of CO_2 separation; Case 2-1 is the 2-stage cascade with 70% degree of CO_2 separation and Case 2-2 with 90% of CO_2 separation. A further comparison of these two concepts with different levels of CO_2 separation was developed to a bar chart shown in Fig. **11**.

From these results it is known that using the membrane cascade for 70% degree of CO_2 separation (Case 1-1) and using the 2-stage cascade for 90% degree of CO_2 separation (Case 2-2) show the optimum values of the capture cost in comparison with the other cases, 31 to 37 euro/t$_{CO2}$/year, respectively. This offers a very interesting hint that for membrane capture system, considering the different separation targets, *e.g.* different levels of CO_2 separation, different system designs should be considered. For the cascade system, the cost of membranes and their frame dominates the whole capital cost; for the 2-stage cascade system, the cost of compression machines (compressor, especially vacuum pump) dominates the whole capital cost. The different depreciation factors are considered for membranes and compression machines (0.064 and 0.225, respectively), owing to the life-time difference. This is reflected to the minor difference between the specific capture cost of the Cases 1-1 and 2-2.

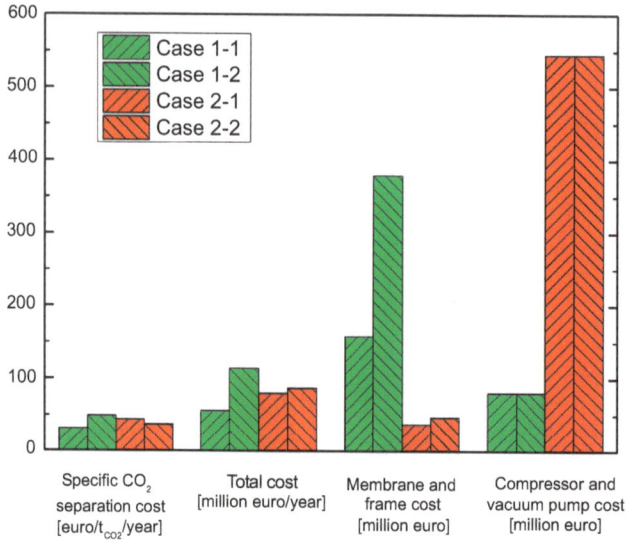

Figure 11: Cost comparison between different cases; Case 1: cascade system (Fig. **5**), 1-1: 70% degree of CO$_2$ separation; 1-2: 90% degree of CO$_2$ separation; Case 2: 2-stage cascade system (Fig. **8**), 2-1: 70% degree of CO$_2$ separation; 2-2: 90% degree of CO$_2$ separation.

4.2. Capture Cost as a Function of Membrane Permeability

The most attractive point of the whole capture cost simulation should be how the membrane parameters (permeability and selectivity) influence the capture cost and the energy consumption. According to the simulation results of the single-stage membrane, the required membrane area is dominated by the CO$_2$ permeance of the membrane. The influence can be clearly seen in Figs. **12a** and **12b**. By increasing the permeability of the membrane, the required membrane area is distinctly decreased, so that the capture cost can be relatively reduced.

For both systems it can be observed that the capture cost can be intensively decreased when the CO$_2$ permeance is increased from 0.5 to 3 Nm^{3}m^{-2}h^{-1}bar^{-1}; but the increasing tendency is sharply slowed down when the CO$_2$ permeance is larger than 3 Nm^{3}m^{-2}h^{-1}bar^{-1}.

As pointed out in the previous paper [42], the energy consumption of the membrane capture system is strongly influenced by the CO$_2$/N$_2$ selectivity, but almost keeps constant by changing CO$_2$ permeance. Fig. **13a** and **13b** show the correlation between the efficiency loss and CO$_2$/N$_2$ selectivity.

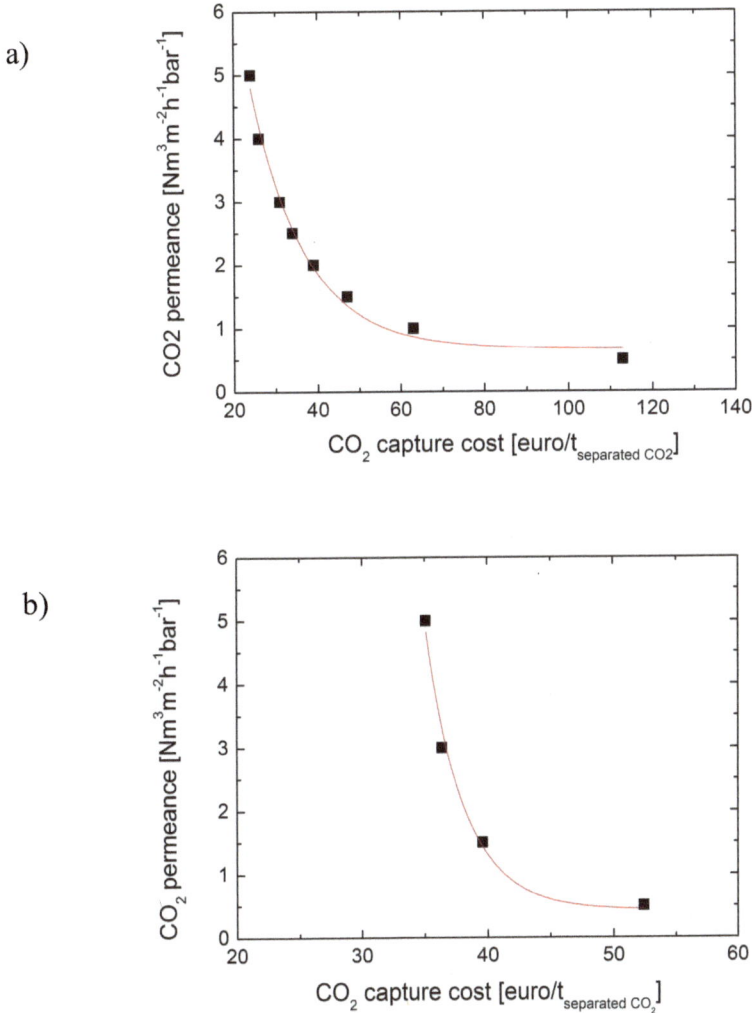

Figure 12: Influence of the CO_2 permeance on the capture cost, CO_2/N_2 selectivity keeps 50; the exponential curve is fitted by the first order; a) The cascade system with 70% degree of CO_2 separation; b) The 2-stage cascade system with 90% degree of CO_2 separation.

It can be observed in Fig. **13**: when the CO_2/N_2 selectivity is increased from 20 to 60, it shows a quite strong decreasing tendency of the efficiency loss; but this tendency is obviously slow when the CO_2/N_2 selectivity is larger than 60. On the basis of the above simulation results, a band width can be predicted for membrane development: CO_2 permeance should be larger than 2 $Nm^3m^{-2}h^{-1}bar^{-1}$, but unnecessary above 10; CO_2/N_2 selectivity should be in the range of 40 to 80. This can lead to a quite promising membrane capture system design. A similar conclusion was obtained in the paper [38] also illustrated in Fig. **14**.

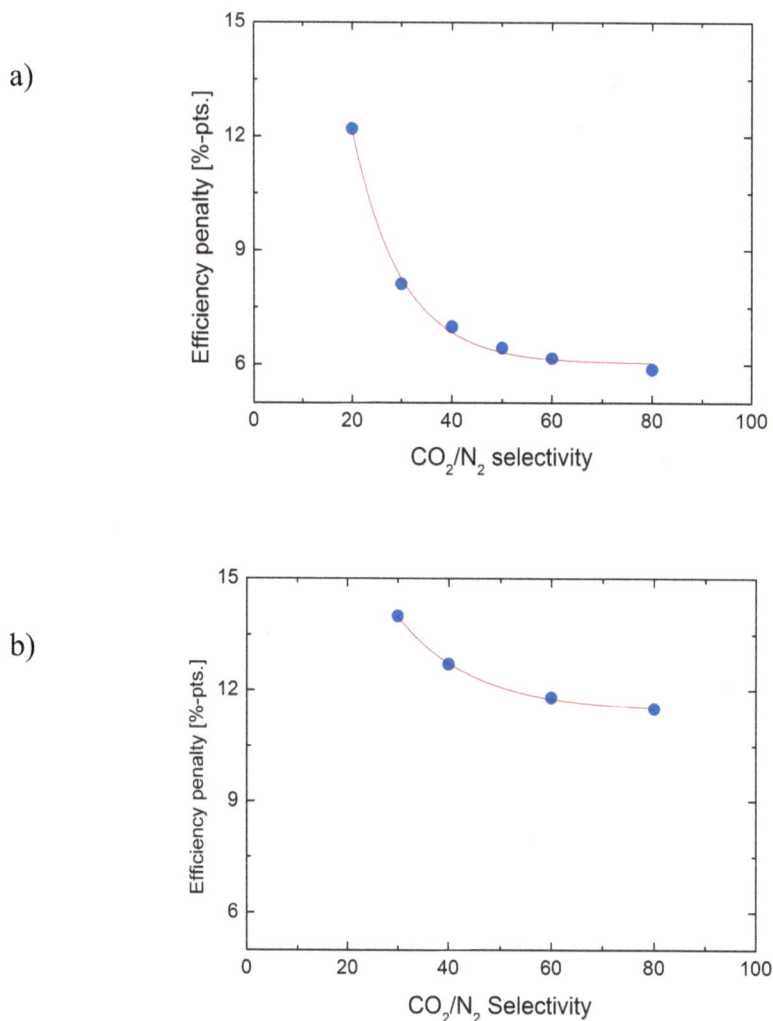

a)

b)

Figure 13: Influence of the CO_2 permeance on the capture cost, CO_2 permeance keeps 3 $Nm^3m^{-2}h^{-1}bar^{-1}$; the exponential curve is fitted by the first order; a) The cascade system with 70% degree of CO_2 separation; b) The 2-stage cascade system with 90% degree of CO_2 separation

Then several pair wise concepts can be coupled together: membrane selectivity and permeability, which decide CO_2 purity and level of CO_2 separation in a single-stage membrane, influence the energy consumption and total membrane area for the multi-stage membrane system, and strongly influence energy and capital cost, respectively.

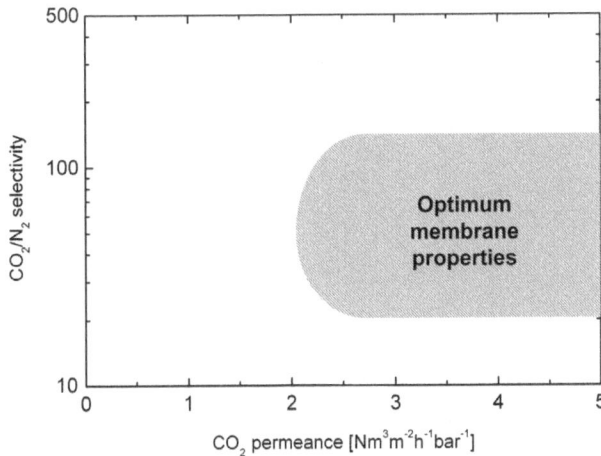

Figure 14: Optimum membrane properties region [38].

CONCLUSION

- Two membrane capture systems (cascade and 2-stage cascade) were investigated in detail, both for energy consumption and capture cost. The results offer us a very interesting hint that for membrane capture system, considering the different separation targets, *e.g.* different levels of CO_2 separation, different system design should be considered.

- Using the NRW reference power plant and based on an ideal feed gas composition of 14 mol% CO_2 and 86 mol% N_2, the cascade system with 70% of CO_2 separation results an energy penalty of 6.4%-points, and the 2-stage cascade system with 90% of CO_2 separation leads to an energy penalty of 12.3%-points, both including CO_2 compression process (110 bar, 30 °C).

- An important feature of three-end membrane capture process is, for 50% and 70% degree of CO_2 separation the energy consumption is lower than that of MEA absorption capture. This shows an application potential for membrane capture as a retrofit for existing power plants.

- Using sweep gas on the permeate side could be an energy-saving concept, but it consumes still energy to separate the captured CO_2 and sweep gas. Furthermore, using sweep gas means that the membrane process must be integrated with some power plant components. The add-on feature of three-end membrane capture system will be lost. 90% of CO_2 separation is a big challenge for any concept, which leads to 12~14%-points efficiency loss.

- Membrane selectivity and permeability decide the CO_2 purity and the level of CO_2 separation in a single-stage membrane, respectively, and strongly influence the energy consumption and total membrane area for a multi-stage membrane system, concerning the energy cost and capital cost respectively. There is a clear trade-off balance between these pair wise parameters.

- A band width can be predicted for multi-stage membrane development: CO_2 permeance should be larger than $2 \ Nm^3m^{-2}h^{-1}bar^{-1}$, but unnecessary above 10; CO_2/N_2 selectivity should lie in the range of 40 to 80. This can lead to a quite promising membrane capture system design for 70% of CO_2 separation.

ACKNOWLEDGEMENTS

Financial support from the Helmholtz Association of German Research Centres (Initiative and Networking Fund) through the Helmholtz Alliance MEM-BRAIN is gratefully acknowledged.

CONFLICT OF INTEREST

The authors confirm that this chapter contents have no conflict of interest.

NOMENCLATURE

a	Depreciation factor (25 years)	
a_m	Depreciation factor (5 years), real interest rate 5%	
A	Membrane area	[m²]
C_{cap}	Capital cost	[euro/a]
$\dot{C}_{CO_2}$	Specific CO_2 separation cost	[euro/t separated CO2/a]
C_{en}	Energy cost per year	[euro/a]
C_{he}	Heat exchangers and cooling facility cost	[million euro]
$C_{O\&M}$	O&M cost	[euro/a]
C_{tot}	Total capture cost	[euro/a]
F_h	Cost factor for housing, installation *etc.*	[-]
K_c	Compressor cost	[million euro]
K_{el}	Electricity cost (hard coal)	[cent/kWh]
K_{ex}	Expander cost	[euro/watt]
K_m	Membrane unit cost	[euro/m²]
K_{mf}	Permanent membrane frame cost	[million euro]
K_{vp}	Vacuum pump cost	[million euro]
I	Investments	[euro]
I_c	Investments for compressor	[euro]
I_{ex}	Investment for expander	[euro]
I_{he}	Investment for heat exchanger	[euro]
I_m	Investment for membrane	[euro]
I_{mf}	Investment for membrane frame	[euro]
I_{vp}	Investment for vacuum pump	[euro]

$M_{CO_2,ann,separated}$	Captured CO_2 per year	[ton]
P_c	Energy used for compressors	[MW]
P_{ex}	Energy recovered by expanders	[MW]
P_{tot}	Total energy consumption	[MW]
P_{vp}	Energy used for vacuum pump	[MW]
t_{op}	Operation time	[h]

REFERENCES

[1] B. Metz, O. Davidson, H. de Coninck, M. Loos, L. Meyer, IPCC Special Report on Carbon Dioxide Capture and Storage, Cambridge University Press, United Kingdom & New York, USA, available in full at www.ipcc.ch, **2005**.

[2] CO_2 capture ready plants, *IEA Greenhouse Gas R&D Programme* (IEA GHG), 2007/4, May **2007**.

[3] The Future of Coal, An Interdisciplinary MIT Study, http://web.mit.edu/coal/The_Future_of_Coal.pdf, **2007**.

[4] Scrubbing CO_2 from Power Plant Flue Gas Using Monoethanolamine (MEA), http://www.netl.doe.gov/newsroom/backgrounder/mb-0002.html.

[5] S. Santos, CO_2 Capture, processing and transport, *IEA Greenhouse Gas R&D Programme*, Asia-Pacific Economic Cooperation (APEC), Mexico City, 24th May, **2007**.

[6] A. Schreiber, P. Zapp, W. Kuckshinrichs, *Environmental impacts of coal-fired power generation with amine-based carbon capture - A life cycle approach,* Proceedings of the 4th European Congress on Economics and Management of Energy in Industry, Porto, 27-30 November, **2007**.

[7] G. Göttlicher, The energetics of carbon dioxide capture in power plants, D.O.E. U.S., available in full at www.netl.doe.gov/publications/carbon_seq/refshelf.html (Ed.), National Energy Technology Laboratory (NETL), **2004**.

[8] H.J. Herzog, What future for carbon capture and sequestration?, *Environ. Sci. Technol.*, **2001**, 35, 148-153.

[9] W.J. Koros, Evolving beyond the thermal age of separation processes: membranes can lead the way, *AICHE J.*, **2004**, 50, 2326-2334.

[10] C.E. Powell, G.G. Qiao, Polymeric CO_2/N_2 gas separation membranes for the capture of carbon dioxide from power plant flue gases, *J. Membr. Sci.*, **2006**, 279, 1-49.

[11] S. Kulprathipanja, Mixed matrix membrane development, *Annals New York Acad. of Sciences*, **2003**, 984, 361.

[12] T.A. Centeno, A.B. Fuertes, Carbon molecular sieve membranes derived from a phenolic resin supported on porous ceramic tubes, *Separation and Purification Technology*, **2001**, 25, 379-384.

[13] T.A. Centeno, J.L. Vilas, A.B. Fuertes, Effects of phenolic resin pyrolysis conditions on carbon membrane performance for gas separation, *J. Membr. Sci.*, **2004**, 270, 101-107.

[14] J.Y.S. Lin, S. Chung, D. Li, J. Ida, J. Park, Dual-phase inorganic membrane for high temperature CO$_2$ separation, http://www.netl.doe.gov/publications/proceedings/04/UCR-HBCU/presentations/Lin_P.pdf.

[15] Ceramic membrane to combat global warming, *Membrane Technology*, **1997**, 92, 11-12.

[16] K. Aoki, K. Kusakabe, S. Morooka, Gas separation properties of A-Type zeolite membrane formed on porous substrate by hydrothermal synthesis, *J. Membr. Sci.*, **1998**, 141, 197-205.

[17] Q. Hu, E. Marand, S. Dhingra, D. Fritsch, W. Wen, G. Wilkes, Poly(amide-imide)/TiO$_2$ nano-composite gas separation membranes: Fabrication and characterization, *J. Membr. Sci.*, **1997**, 135, 65-79.

[18] P. Kumar, J. Ida, V.V. Guliants, High flux mesoporous MCM-48 membranes: Effects of support and synthesis conditions on membrane permeance and quality, *Micropor. and Mesopor. Mater.*, **2007**, 5, doi:10.1016/j.micromeso.2007.1006.1040.

[19] L.M. Robeson, Correlation of separation factor *versus* permeability for polymeric membranes, *J. Membr. Sci.*, **1991**, 62, 165-185.

[20] B.D. Freeman, Basis of permeability/selectivity tradeoff relations in polymeric gas separation membranes, *Macromolecules*, **1999**, 32, 375-380.

[21] E. Favre, Carbon dioxide recovery from post-combustion processes: Can gas permeation membranes compete with absorption? *J. Membr. Sci.*, **2007**, 294, 50-59.

[22] A. Brunetti, F. Scura, G. Barbieri, E. Drioli, Membrane technologies for CO$_2$ separation, *J. Membr. Sci.*, **2010**, 359, 115-125.

[23] D. Shekhawat, D.R. Luebke, H.W. Pennline, A review of carbon dioxide selective membranes, NETL & DOE, 1st December, **2003** (http://www.osti.gov/bridge/product.biblio.jsp.osti_id=819990).

[24] M. Langsam, Polyimides for gas separation, *Plastics Engineering*, **1996**.

[25] T. Kouketsu, S. Duan, T. Kai, S. Kazama, K. Yamada, PAMAM dendrimer composite membrane for CO$_2$ separation: Formation of a chitosan gutter layer, *J. Membr. Sci.*, **2007**, 287, 51-59.

[26] L. Deng, T.-J. Kim, M.-B. Hägg, Facilitated transport of CO$_2$ in novel PVAm/PVA blend membrane, *J. Membr. Sci.*, **2008**, 340, 154-163.

[27] R. Mahajan, W.J. Koros, Factors controlling successful formation of mixedmatrix gas separation materials, *Industrial Engineering and Chemistry Research*, **2000**, 39, 2692-2696.

[28] C.M. Zimmerman, A. Singh, W.J. Koros, Tailoring mixed matrix compositemembranes for gas separations, *J. Membr. Sci.*, **1997**, 137, 145-154.

[29] C.W. Jones, W.J. Koros, Carbon molecular sieve gas separation membranes-II. Regeneration following organic exposure, *Carbon*, **1994**, 32, 1427-1432.

[30] C.W. Jones, W.J. Koros, Carbon molecular sieve gas separation membranes-I. Preparation and characterization based on polyimide precursors, *Carbon*, **1994**, 32, 1419-1425.

[31] S.M. Saufi, A.F. Ismail, Fabrication of carbon membranes for gas separation - a review, *Carbon*, **2004**, 42, 241-259.

[32] A. Car, C. Stropnik, W. Yave, K.V. Peinemann, Pebax/polyethylene glycol blend thin film composite membranes for CO$_2$ separation, *Separation and Purification Technology*, **2008**, 62, 110-117.

[33] A. Car, C. Stropnik, W. Yave, K.V. Peinemann, PEG modified poly(amide-*b*-ethylene oxide) membranes for CO$_2$ separation, *J. Membr. Sci.*, **2008**, 307, 88-95.

[34] A. Car, C. Stropnik, W. Yave, K.V. Peinemann, Tailor-made polymeric membranes based on segmented block copolymers for CO$_2$ separation, *Advanced Functional Materials*, **2008**, 18, 2815-2823.

[35] W. Yave, A. Car, S.S. Funari, S. Nunes, K.V. Peinemann, CO$_2$-Philic Polymer Membrane with Extremely High Separation Performance, *Macromolecules*, **2010**, 43, 326-333.

[36] W. Yave, A. Car, K.V. Peinemann, Nanostructured membrane material designed for carbon dioxide separation, *J. Membr. Sci.*, **2010**, doi:10.1016/j.memsci.2009.12.019.

[37] A. Hussain, M.-B. Hägg, A feasibility study of CO$_2$ capture from flue gas by a facilitated transport membrane, *J. Membr. Sci.*, **2010**, 359, 140-148.

[38] T.C. Merkel, H. Lin, X. Wei, R. Baker, Power plant post-combustion carbon dioxide capture: An opportunity for membranes, *J. Membr. Sci.*, **2010**, 359, 126-139.

[39] Konzeptstudie: Referenzkraftwerk Nordrhein-Westfalen (RKW NRW), *VGB Power Tech e.V.*, Essen, Germany, February **2004**.

[40] L. Zhao, E. Riensche, R. Menzer, L. Blum, D. Stolten, A parametric study of CO$_2$/N$_2$ gas separation membrane processes for post-combustion capture, *J. Membr. Sci.*, **2008**, 325, 284-294.

[41] E. deVisser, C. Hendriks, M. Barrio, M.J. MØlnvik, G. deKoeijer, S. Liljemark, Y. LeGallo, Dynamis CO$_2$ quality recommendations, *International Journal of Greenhouse Gas Control*, **2008**, 2, 478-484.

[42] L. Zhao, E. Riensche, L. Blum, D. Stolten, Multi-stage gas separation membrane processes with post-combustion capture: energetic and economic analyses, *J. Membr. Sci.*, **2010**, 359, 160-172.

[43] L. Zhao, R. Menzer, E. Riensche, L. Blum, D. Stolten, Concepts and investment cost analyses of multi-stage membrane systems for carbon dioxide recovery in post-combustion processes, *Energy Procedia*, **2009**, 1, 269-278.

[44] W. Arlt, Thermodynamical optimization of solvents for CO$_2$ absorption, Workshop CO$_2$-Capture, -Utilization and -Sequestration, Status and Perspectives, PROCESSNET, DECHEMA-Haus, Frankfurt am Main, Germany, 21-22 January, **2008**.

[45] P. Galindo-Cifre, K. Brechtel, S. Hoch, H. García, N. Asprion, H. Hasse, G. Scheffknecht, Integration of a chemical process model in a power plant modelling tool for the simulation of an amine based CO$_2$ scrubber, *Fuel*, **2009**, 88, 2481-2488.

[46] M.R.M. Abu-Zahra, J.P.M. Niederer, P.H.M. Feron, CO$_2$ capture from power plants Part II. A parametric study of the economical performance based on mono-ethanolamine, *International Journal of Greenhouse Gas Control* I, **2007**, 135-142.

[47] C. Hendriks, *Carbon dioxide removal from coal-fired power plants*, Kluwer Academic Publishers, Dordrecht/Boston/London, **1994**.

[48] D. Singh, E. Croiset, P.L. Douglas, M.A. Douglas, Techno-economic study of CO$_2$ capture from an existing coal-fired power plant: MEA scrubbing *vs.* O$_2$/CO$_2$ recycle combustion, *Energy Convers. Manage.*, **2003**, 44, 3073-3091.

[49] J.P. van der Sluijs, C.A. Hendriks, K. Blok, Feasibility of polymer membranes for carbon dioxide recovery from flue gases, *Energy Convers. Manage.*, **1992**, 33, 429-436.

[50] MEM-BRAIN Alliance (Gas separation membranes for zero-emission fossil power plants), 18 research institutions and 5 industrial partners Coordinator: IEF-1, Forschungszentrum Jülich GmbH, Oct. 2007 - Jun. **2011**.

[51] P.H.M. Feron, C.A. Hendriks, CO$_2$ capture process principles and costs, *Oil & Gas Science and Technology* - Rev. IFP, **2005**, 60, 451-459.

CHAPTER 3

Minimizing Energy Consumption in CO_2 Capture Processes Through Process Integration

Alma Esthela Torres-López[1], Martín Picón-Núñez[1,*] and Rosa-Hilda Chavez[2]

[1]Department of Chemical Engineering, University of Guanajuato, Guanajuato, Gto.36050, México and [2]National Institute of Nuclear Research, Carretera Mexico-Toluca S/N, La Marquesa, Ocoyoacac, 52750, Mexico

Abstract: This chapter reviews the thermal integration for minimum external energy consumption of CO_2 capture process using amines. Post combustion capture of CO_2 through the use of amines is a well established technique; the stand alone process is highly energy-intensive since the recovery of the amine solution is achieved through the use of a separation process where heating and cooling are required. Energy integration of the hot flue gas coming from a power station plant from where CO_2 is absorbed can serve the purpose of providing the heating and cooling needs of the process. Heat recovery through steam raising is considered for heating, cooling and for the production of power for the operation of pumps and compressors. The results show that the needs of the largest energy user of the process can be fully met by heat integration.

Keywords: Minimum energy consumption, CO_2 capture, process integration, numerical simulation.

1. INTRODUCTION

Interest in the recovery of CO_2 from flue gases from power plants responds to two fundamental issues: one is the reduction of the environmental impact of CO_2 in the atmosphere and the other is its commercial use. Since power plants are sources of CO_2 production in large quantities, its recovery in this type of processes represents a feasible option; however, care must be taken that the separation process does not consume more energy so that more CO_2 gets generated elsewhere. One way of ensuring minimum energy consumption in processes is maximizing the energy recovery. The process integration techniques (such as

---

*Corresponding author Martín Picón-Núñez: Department of Chemical Engineering, University of Guanajuato, Noria Alta S/n, Guanajuato, Gto., C.P. 36050, México; Tel: +52; E-mail: piconnunez@gmail.com

pinch technology) offer the most practical means of recovering energy at minimum capital cost.

Among the post combustion CO_2 recovery techniques is the absorption using chemical solvents, where the flue gas is put into contact with a solvent. In this work, the absorption using monoethanolamine (MEA) is analyzed. MEA shows a high selectivity level compared to other solvents for the recovery of CO_2. However, the process consumes a large amount of energy, particularly during the solvent regeneration stage.

The recovery of CO_2 was initiated in 1970, not because of environmental concerns but because of its economical importance in oil extraction. About 80% of CO_2 production is currently used for this purpose [1]. CO_2 capture has been implemented in small power plants but not in large ones which becomes a challenge for future applications. A majority of the existing CO_2 absorption processes use MEA as solvent; the main commercial processes available are: the Fluor Daniel Econamine FG process and the ABB Lummus Crest MEA process [2, 3]. The Econamine FG process uses 30% wt MEA solution and oxygen as inhibitor for reducing the solvent degradation and to avoid corrosion of the equipment. The ABB Lummus process uses an MEA solution between 15% and 20% wt without inhibitor [4]. Researches in the field have reviewed the modeling of process [5-9] and others have focused on the reduction of the energy requirements. Studies have been conducted that seek to minimize the amount of thermal (heating and cooling) demand and the electric energy (compression and pumping) required for the operation of the CO_2 absorption process. The use of pinch technology to reduce the energy consumption in a CO_2 absorption system from a coal fired power plant has been analyzed by some authors [10]. A model has been proposed for exergy recovery in an advanced integrated gasification-combined cycle; this model achieves higher plant efficiency by applying an autothermal reaction in the gasifier [11]. Such approach considers an additional heat supply from the gas turbine exhaust and the steam extracted from the steam turbine. Studies have been performed where process modifications for reducing energy consumption have been assessed and it has been demonstrated that the lowest energy requirement results when the modification takes place when the absorber operates with a stripper lean vapor recompression system [12]. In this chapter we look at application of traditional thermal integration techniques that seek to maximize the heat recovery from the hot flue gases to produce medium pressure steam which can be used to run the re-boiler of the absorber, pumps and compressors and even absorption refrigeration systems.

2. CO_2 RECOVERY TECHNIQUES

The techniques used for the separation of CO_2 from combustion gases are based on physical and chemical processes such as: absorption, adsorption, membranes and cryogenic processes as shown in Fig. **1**. The selection of the proper technology depends on various aspects such as: concentration of CO_2 in the gas stream, types of contaminants present, *etc.*, [13, 14]. Many of the various technologies for CO_2 separation cannot be applied to large scale generation [15]. The existing technologies for CO_2 recovery can be classified into three large groups:

2.1. Pre-Combustion

In this technique the primary fuel is treated in order to produce separate streams of CO and H_2. The carbon content of the fuel is captured before combustion in a gasifier that produces CO which is later reacted with water to obtain a rich CO_2 stream [16].

2.2. Combustion of Oxi-Fuel

In this technique combustion air is replaced with pure oxygen in order to increase the CO_2 concentration in the flue gas.

2.3. Post-Combustion

In this technique CO_2 capture takes place after combustion and imposes no changes to the power plant since the recovery is carried out directly from the flue gas stream.

3. CO_2 ABSORPTION PROCESS DESCRIPTION

Chemical absorption processes take advantage of the reversible nature of the absorption-desorption of CO_2 from an aqueous alkaline solution. Fig. **2** shows a flow sheet of the CO$_2$ recovery process using MEA. The CO$_2$ gas stream at approximately 150 °C is cooled down to reduce its temperature to 40 °C. Then it is passed through an absorption column where it is put into contact with the absorbing solution. The CO$_2$ contained in the gas stream reacts with the solvent and then absorbed in it. The CO$_2$ rich stream is recovered from the bottom of the column and is then passed through a heat recovery exchanger where it is preheated before entering a second column where CO$_2$ is desorbed by the addition of heat. The CO$_2$ is then released and compressed for its transportation [16, 17]. The recovered amine solution obtained from the bottom of the stripping column is cooled down in a heat recovery exchanger, and then it is mixed with fresh MEA solution before being sent back to the absorption column. The main parts of equipment of the flow sheet in Fig. **2** are: P01, a water pump; C01, a gas compressor; V01, a direct contact gas cooler; T01, a CO$_2$ absorption column; P02, a pump; E01, a heat exchanger; T02, a CO$_2$ stripping column; T02-C, the condenser of the stripping column; T02-R, the re-boiler of the stripping column; E02, a cooler and V02, a stream mixer.

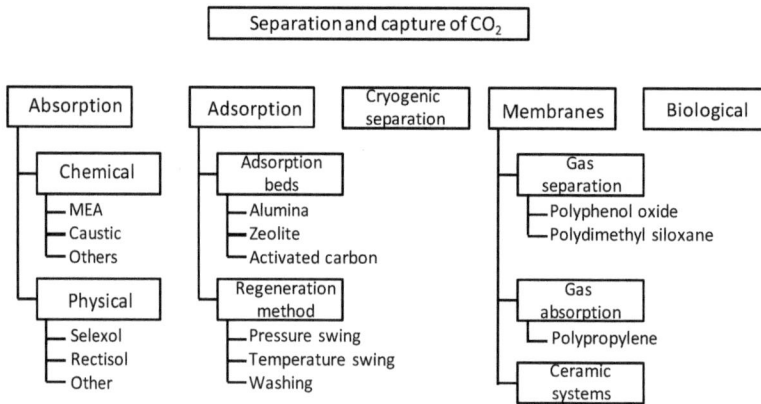

Figure1: Available techniques for the separation and sequestration of CO$_2$.

The main organic solvents of commercial interest for CO$_2$ purification are amines. There are various types, for instance: primary amines such as monoethanolamine (MEA), secondary amines such as diethanolamine (DEA) and ternary amines such as methildiethanolamine (MDEA). Primary amines are more reactive with carbon and are therefore superior compared to others. Secondary and ternary amines exhibit a large reactive capacity but the reaction velocity is too slow. Inorganic solvents such as potassium carbonate, sodium carbonate and ammonia aqueous solutions are also used.

Process simulation of the flow sheet in Fig. **2** has been conducted using a commercial process simulator [18]. There are two important parameters to consider for a maximum CO$_2$ recovery, these are: the molar ratio of MEA to CO$_2$ (α), and the MEA concentration.

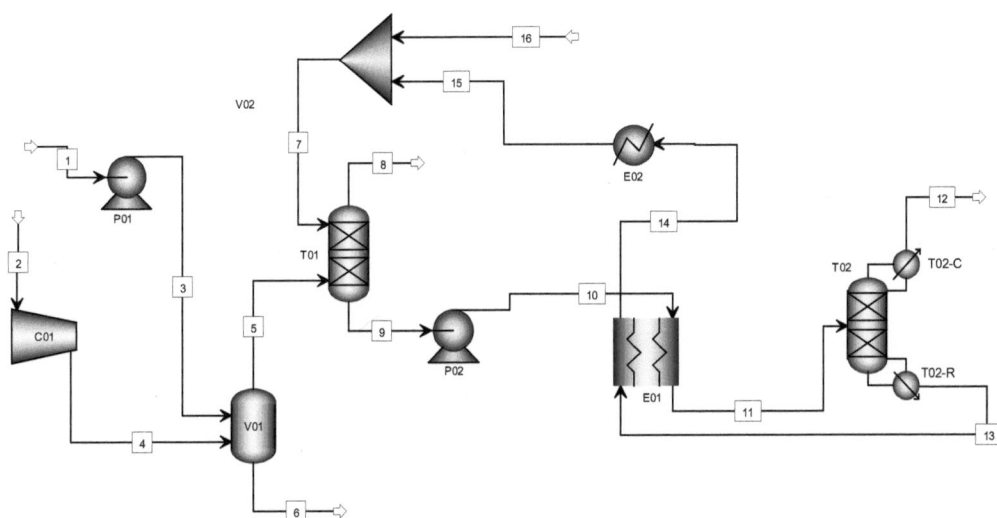

Figure 2: Flow sheet of the CO$_2$ absorption process with MEA.

The amount of CO$_2$ removed from the combustion gases is strongly affected by the mass transfer process. The efficiency of the CO$_2$ removal process in the absorber is a function of various parameters which affect the liquid-gas equilibrium such as: mass flow rate, gas composition, temperature, pressure and concentration of MEA.

The thermodynamics of the system is better described by the Electrolyte-NRTL model embedded in Aspen Plus. Several chemical reactions take place during the absorption and regeneration stages of the process. The dissociation and precipitation processes are too fast that the reactions can be considered to be at equilibrium. These types of reactions in liquid phase are called chemical solution and can be represented by means of the following equilibrium reactions [13]:

$$2H_2O \rightleftharpoons OH^- + H_3O^+$$

$$CO_2 + 2H_2O \rightleftharpoons HCO_3^- + H_3O^+$$

$$HCO_3^- + H_2O \rightleftharpoons CO_3^{2-} + H_3O^+$$

$$RNH_3^+ + H_2O \rightleftharpoons RNH_2^+ + H_3O^+$$

$$RNHCOO^- + H_2O \rightleftharpoons RNH_2^+ + HCO_3^-$$

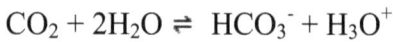

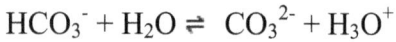

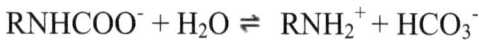

There are some models for the determination of the physical properties of some of the compounds present in the CO_2-MEA-H_2O process. However, the electrolyte-NRTL is the one that best suits the system. It is an extended method that incorporates the interactions between ions in solution. For the determination of the physical properties, the method uses binary interaction parameters and equilibrium constants for systems containing: CO_2, H_2S, MEA and H_2O for temperatures up to 120 °C and concentrations up to 50%wt. Other methods available in the simulator are: emea, kemea, mea, kmea and amine.

For the simulation, the following assumptions are made: a) the process consists of two stages, namely: absorption and desorption, b) monoethanolamine in aqueous solution is used, c) the flue gas is free from contaminants such as: NOx and SOx, d) the separation columns are modeled using the RadFrac module and they operate at a constant pressure, e) the reaction between CO_2 and MEA is fast enough to be modeled as an equilibrium reaction, f) the initializing values are taken from previous works and are shown in Table **1**. The flue gas and solvent compositions are given in Table **2**. Amine concentration is varied in a range between 10 and 45% wt; similarly, the molar ratio α (mol CO_2/mol MEA) is varied between 0.2 and 0.6.

Table 1: Values used to initialize the simulation [13]

Initial Considerations			
MEA Temperature	40 °C	**Flue gas mass flow rate**	19.41102 kg/s
MEA Inlet pressure	110 k Pa	**Flue gas temperature**	40 °C
No. of stages in absorber	10	**Flue gas pressure**	120 k Pa
No. of stages in stripper	10	**MEA concentration**	30 wt%

The simulation of the process has been developed in order to determine essential operating conditions, mass and energy balances and equipment dimensions. The CO_2 gas stream produced in a 100 MW power plant has been analyzed. Details of design specifications for the simulation modules used are displayed in Table **3**.

Table 2: Flue gas and solvent composition

Flue Gas Composition				
ID	Name	Chemical Formula	Molecular weight	Composition
H₂O	Water	H_2O	18	12.07
CO₂	Carbon dioxide	CO_2	44	11.79
N₂	Nitrogen	N_2	28	72.01
O₂	Oxygen	O_2	32	4.12
Solvent Composition				
H₂O	Water	H_2O	18	90-60
MEA	Monoethanolamine	C_2H_7NO	61	10-40

Fig. **3** shows the effect of the molar ratio CO_2/MEA (α) and MEA concentration upon CO_2 recovery. It can be seen that for a molar ratio of 0.5 the recovery reaches its maximum value. On the other hand, a 10% MA concentration allows for a higher CO_2 recovery. Table **4** summarizes these results.

Table 3: Design specifications of simulation modules

Module	Specification
Absorption	Design mode: stages in equilibrium
	Number of stages: 10
	Operating pressure: 110 k Pa
T01/Rad Frac	Packing type: Raschig
	Packing size: 35mm
	Packing height per stage: 0.916m
Desorption	Design mode: stages in equilibrium
	Number of stages: 10
	Operating pressure: 115 k Pa
T02/Rad Frac	Packing type: Raschig
	Packing size: 50mm
	Packing height per stage: 0.916 m

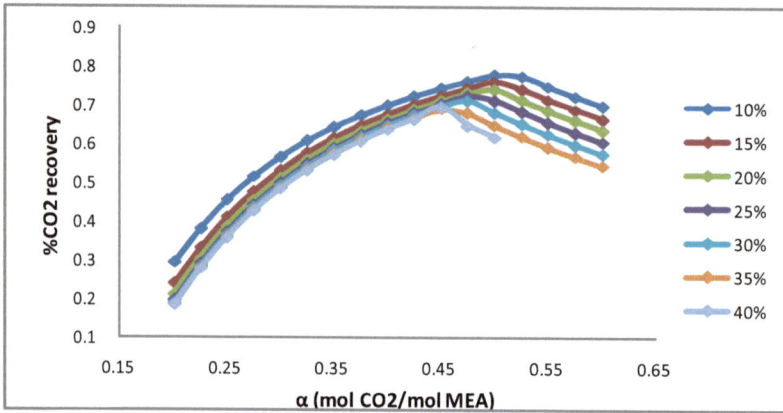

Figure 3: Effect of the molar ratio α upon the CO_2 recovery for different MEA concentrations.

The energy and mass balances for the absorption of CO_2 with a value of α of 0.5 and 10% MEA concentration are shown in Table **5**. The energy requirements of each piece of equipment are presented in Table **6**. The diameters of the absorption and desorption columns for the various scenarios of MEA concentrations are presented in Table **7**.

Table 4: CO_2 recovery for various MEA concentrations

MEA concentration	10%	15%	20%	25%	30%	35%
CO$_2$% recovery	73.07	71.15	67.3	65.4	61.4	59.62

Table 5: Process mass balance for α=0.5 and a 10% MEA concentration

COMPONENTS k mol/s	STREAM							
	1	2	3	4	5	6	7	8
H$_2$O	1.388	0.130	1.388	0.130	0.038	1.479	3.279	0.087
MEA	0	0	0	0	0	0	0.102	1.49E-05
H$_2$S	0	0	0	0	0	0	0	0
CO$_2$	0	0.052	0	0.052	0.052	6.42E-05	0	1.46E-10
HCO$_3$-	0	0	0	0	0	9.30E-07	0	0
MEACOO-	0	0	0	0	0	0	0	0
MEA+	0	0	0	0	0	0	0.002	0

Table 5: contd….

CO_3^{-2}	0	0	0	0	0	1.63E-12	0	0
HS-	0	0	0	0	0	0	0	0
S_{-2}	0	0	0	0	0	0	0	0
H_3O^+	2.51E-09	0	2.51E-09	0	0	9.30E-07	1.82E-13	0
OH^-	2.51E-09	0	2.51E-09	0	0	2.22E-11	0.002	0
N_2	0	0.499	0	0.499	0.498	0.002	0	0.495
O_2	0	0.025	0	0.025	0.025	2.25E-04	0	0.024
Total Flow rate (k mol/s)	1.388	0.706	1.388	0.706	0.612	1.481	3.385	0.605
Total flow rate (kg/s)	25	19.411	25	19.411	17.705	26.706	65.458	16.193
Temperature (K)	298.2	423.2	298.2	487.4	313.2	313.2	313.2	328.6
Pressure (k Pa)	101	101.3	150	150	120	120	110	110
Enthalpy (kW)	-3.97E+08	-4.91E+07	-3.97E+08	-4.77E+07	-2.94E+07	-4.21E+08	-9.63E+08	-2.04E+07
	STREAM							
COMPONENTS (k mol/s)	9	10	11	12	13	14	15	16
H_2O	3.219	3.220	3.217	0.036	3.194	3.194	3.194	0.083
MEA	0.0133	0.012	0.015	9.19E-25	0.078	0.077	0.076	0.024
H_2S	0	0	0	0	0	0	0	0
CO_2	1.41E-4	5.78E-05	2.01E-04	0.038	7.07E-09	9.53E-07	1.51E-08	0
HCO_3^-	0.013	0.011	0.014	0	0.002	0.001	3.06E-4	0
MEACOO-	0.038	0.039	0.037	0	0.012	0.012	0.012	0
MEA^+	0.053	0.053	0.052	0	0.014	0.015	0.016	0.003
CO_3^{-2}	9.13E-4	1.17E-3	7.35E-4	0	2.83E-4	5.22E-4	1.76E-3	0
HS^-	0	0	0	0	0	0	0	0
S^{-2}	0	0	0	0	0	0	0	0
H_3O^+	8.22E-10	4.77E-10	1.23E-09	0	3.33E-10	1.23E-10	7.66E-12	1.42E-15
OH^-	1.55E-06	1.33E-06	1.64E-06	0	5.16E-05	5.73E-05	7.49E-05	0.003
N_2	0.003	0.003	0.003	0.003	1.53E-20	0	0	0
O_2	3.99E-04	3.98E-4	3.98E-4	3.98E-4	6.19E-19	0	0	0

Table 5: contd….

Total Flow rate (k mol/s)	3.340	3.340	3.340	0.078	3.3	3.3	3.3	0.112
Total flow rate (kg/s)	66.97	66.97	66.97	2.428	64.542	64.542	64.542	3.134
Temperature (K)	320.5	320.6	340	356.1	378.8	359.8	313.2	313.2
Pressure (k Pa)	115	130	125	115	120	120	110	117
Enthalpy (kW)	-9.72E+08	-9.72E+08	-9.67E+08	-2.36E+07	-9.29E+08	-9.34E+08	-9.46E+08	-3.15E+07

Table 6: Energy requirements (kW) for the absorption of CO_2

MEA%	MEA%						
Equipment	10%	15%	20%	25%	30%	35%	40%
P01	1.77	1.77	1.77	1.77	1.77	1.77	1.77
C01	1,405.08	1,405.08	1,405.08	1,405.08	1,405.08	1,405.08	1,405.08
P02	2.11	0.94	0.74	0.65	0.57	0.53	0.48
E01	4,939.62	3,846.97	2,996.46	2,436.10	1,987.40	1,705.12	1,427.45
T02-C	-18,013.89	-18,134.86	-17,808.81	-17,754.05	-17,569.05	-17,491.59	-17,570.62
T02-R	38,237.11	32,339.09	30,178.65	29,302.88	28,333.39	27,892.41	27,535.87
E02	-11,622.85	-6,830.03	-4,834.27	-4,115.53	-3,444.09	-3,238.71	-2,936.14

Table 7: Diameters of the absorption and desorption columns required for the various MEA concentrations

Column Diameter (m)							
Column	MEA%						
	10%	15%	20%	25%	30%	35%	40%
T01	4.78	4.47	4.23	4.22	4.05	4.06	3.97
T02	4.54	4.19	3.98	3.88	3.78	3.74	3.63

4. TARGETING FOR PROCESS HEATING AND COOLING NEEDS

The external energy requirements of any process can be determined once the heat and mass balance are known. Energy requirements take the form of process heating and cooling demands. The best known approach for determining these needs is Pinch Analysis [19]. The concept of the "pinch" from where the various tools for the design of integrated production systems evolved represents the minimum temperature approach for heat recovery and it can graphically be described by means of the composite curves as shown in Fig. **4**. The diagram consists of two separate curves; the hot composite that represents the total amount of heat that must be eliminated from the process, and the cold composite curve that represents the total heating needs of it. Both curves can be superimposed so that they overlap over a range along the enthalpy axis. The extent of the overlap indicates the amount of thermal energy that can be recovered from the hot streams and transferred to the cold streams. Apart from indicating the extent of heat recovery, the composite curves also indicate the amount of external heating and cooling the process requires for completing the energy balance. The relationship between the level of heat recovery and the external consumption of energy is inverse, *i.e.* the larger the amount of heat recovery the lower the need for external energy consumption.

The composite curves also show the temperature driving forces available for heat transfer. The point where the minimum temperature approach occurs between the curves is the pinch and its value can directly be related to the amount of energy recovered, the external energy consumption and the capital investment in terms of heat transfer equipment. Thermodynamically, the pinch divides the process into two regions: above the pinch where the process becomes a heat sink and below the pinch where the process becomes a heat source (Fig. **4**).

The construction of the composite curves is the first stage in the determination of the heating and cooling needs of a process. For its construction, the proper data must be gathered. For new processes, data can be extracted once the process flow sheet and the heat and mass balance for a given throughput have been specified. This data includes the identification of the hot and cold streams. Cold streams are those that must absorb heat in order to increase its temperature to a given target and hot streams are those that require heat to be removed. The basic stream data to be extracted includes: mass flow rates (m), specific heat capacity (C_p), supply temperature (T_s) and target temperature (T_T). Table **8** shows typical stream data for energy integration.

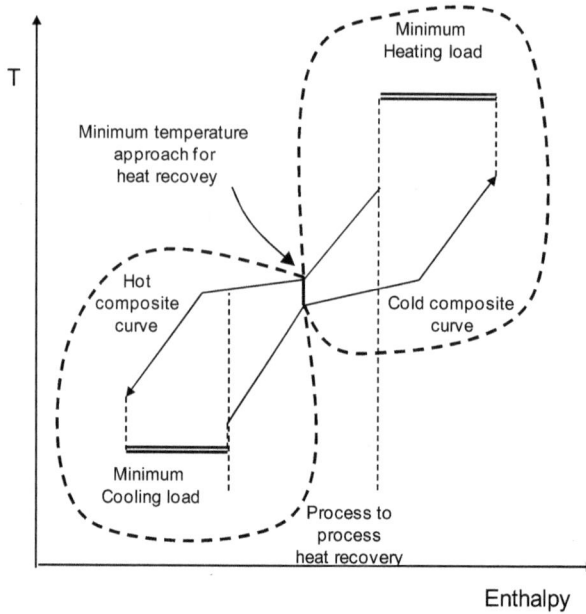

Figure 4: Process composite curves.

When more than one hot and cold stream are present in the process, they can be added together graphically giving rise to the composite curves. The thermal addition of the cold streams results in the cold composite curve. The procedure that describes the way in which two simple hot streams are added together in a temperature *vs.* enthalpy diagram is shown in Fig. **5**. The same approach applies for the case of the cold streams.

Table 8: Process data for energy integration

Stream Type	Name	Ts (°C)	T$_T$ (°C)	mC$_p$ (kW/°C)	ΔH (kW)
Cold	C1	25	200	25	4,375
Cold	C2	90	130	30	2,000
Cold	C3	120	200	40	3,200
Cold	C4	160	160	-	5,000
Hot	H1	200	90	65	7,150
Hot	H2	90	90	-	4,500
Hot	H3	90	40	15	750
Hot	H4	160	80	20	1,200

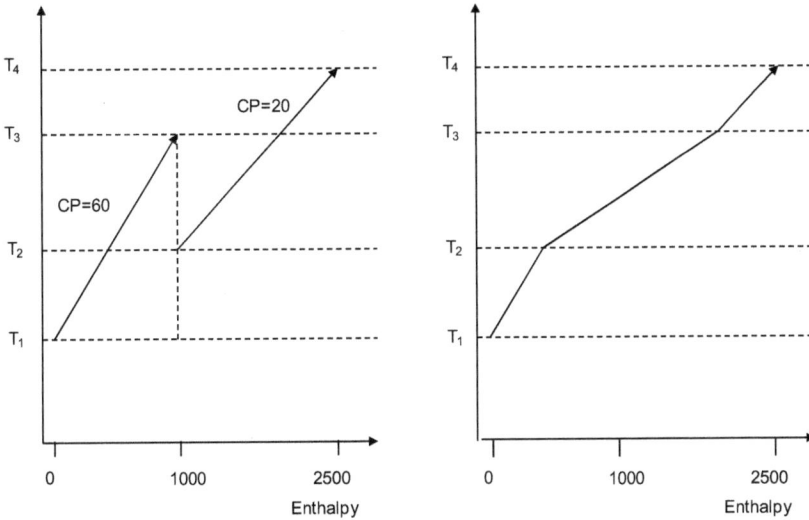

Figure 5: Construction of a hot composite curve.

The heating and cooling external duties of a process depend on the level of heat recovery. The way to calculate the heating and cooling needs for a given ΔT_{min} is through the application of the Problem Table Algorithm that is described below:

1. Determination of modified temperatures. For the specified ΔT_{min}, the supply and target temperatures of each stream are modified by $\frac{1}{2}\Delta T_{min}$. Hot streams are affected by $-\frac{1}{2}\Delta T_{min}$ and cold streams by $+\frac{1}{2}\Delta T_{min}$. The modified temperature data of the case study considering a ΔT_{min} of 12 °C is shown in Table **9**.

Table 9: Modified data for $\Delta T_{min}=12$ °C

Stream type	Name	T_s^* (°C)	T_T^* (°C)	mC_p (kW/ °C)	ΔH (kW)
Cold	C1	31	206	25	4,375
Hot	H1	194	84	65	7,150
Cold	C2	96	136	30	2,000
Cold	C3	126	206	40	3,200
Hot	H2	84	84	-	4,500
Cold	C4	166	166	-	5,000
Hot	H3	84	34	15	750
Hot	H4	154	74	20	1,200

2. Energy balances per temperature interval. For each modified temperature interval, enthalpy balances are carried out in order to determine whether a surplus or deficit of heat exists at the corresponding interval. Fig. **6** shows the enthalpy balances per temperature interval.

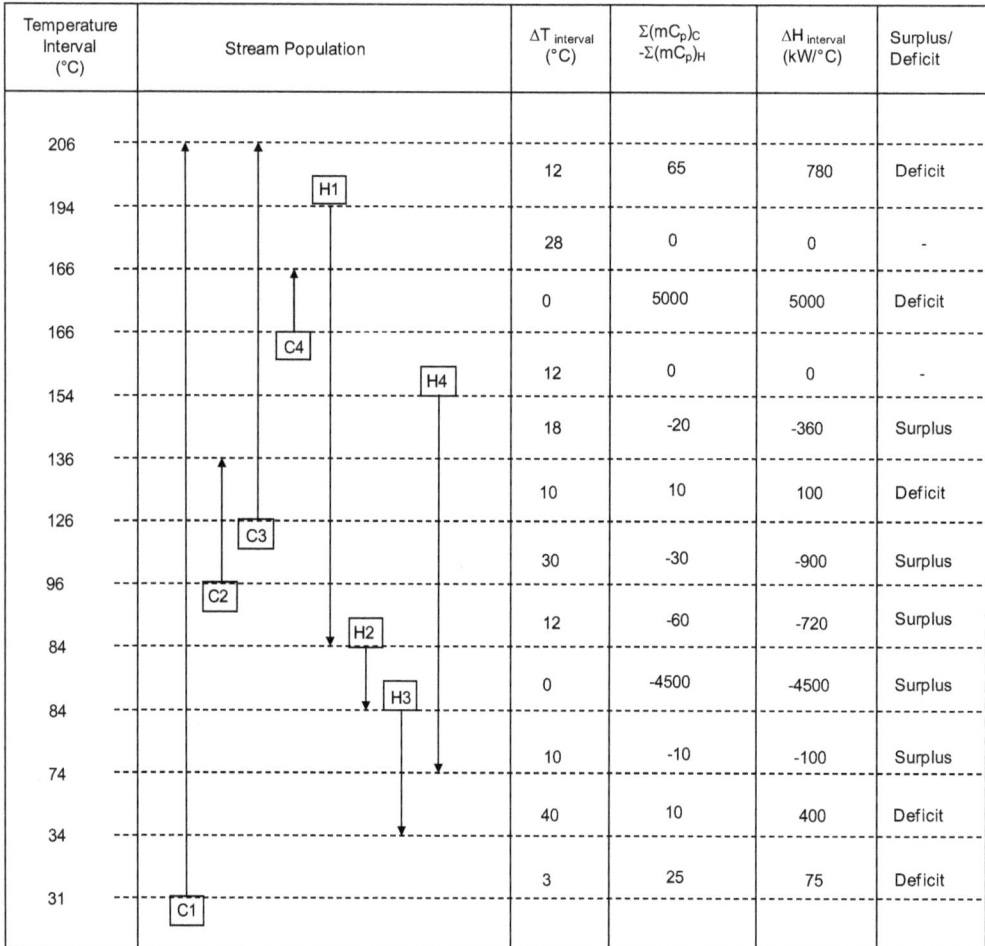

Temperature Interval (°C)	Stream Population	$\Delta T_{interval}$ (°C)	$\Sigma(mC_p)_C$ $-\Sigma(mC_p)_H$	$\Delta H_{interval}$ (kW/°C)	Surplus/ Deficit
206	H1	12	65	780	Deficit
194		28	0	0	-
166	C4	0	5000	5000	Deficit
166	H4	12	0	0	-
154		18	-20	-360	Surplus
136	C3	10	10	100	Deficit
126		30	-30	-900	Surplus
96	C2	12	-60	-720	Surplus
84	H2	0	-4500	-4500	Surplus
84	H3	10	-10	-100	Surplus
74		40	10	400	Deficit
34		3	25	75	Deficit
31	C1				

Figure 6: Energy balances per temperature interval.

3. Heat cascade. The principle behind heat cascading is that any surplus of heat is allowed to flow down to lower temperature intervals to supply the heat need in that interval. A first heat cascade exercise as the one shown in Fig. **7(a)** reveals that the largest heat deficit of -

5,780 units flows across the temperature line representing the modified temperature of 166 °C. This deficit must be compensated for by adding from the amount of heat at the top that makes this deficit zero; so 5,780 of heat are added and the heat cascade process is repeated as shown in Fig. **7(b)**. The point where the heat flow equals zero represents the modified pinch temperature. The amount of heat added at the top is the minimum energy consumption (5,780 kW) and the amount of heat released at the lower temperature is the minimum cooling required by the process (6,005 kW) for the specified minimum temperature approach.

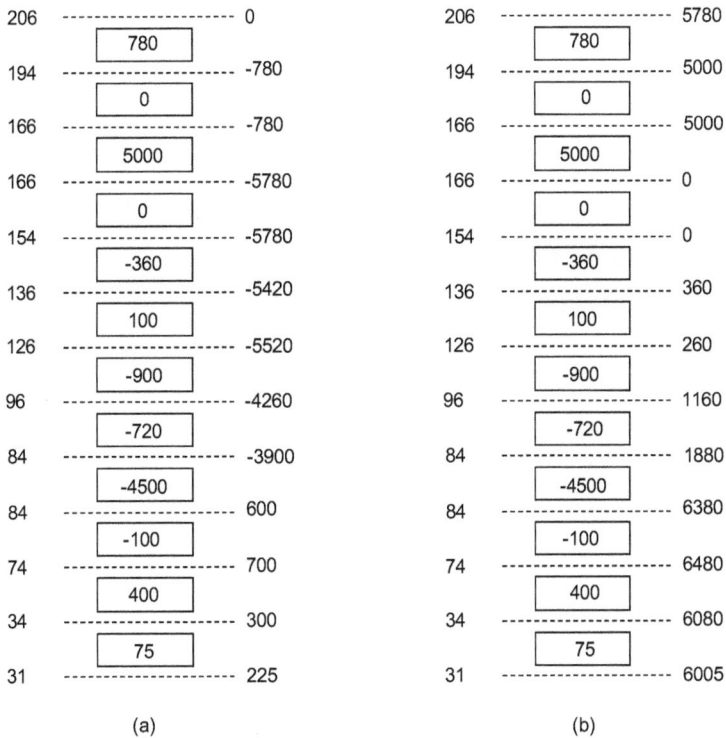

(a)		(b)	
206	0	206	5780
780		780	
194	-780	194	5000
0		0	
166	-780	166	5000
5000		5000	
166	-5780	166	0
0		0	
154	-5780	154	0
-360		-360	
136	-5420	136	360
100		100	
126	-5520	126	260
-900		-900	
96	-4260	96	1160
-720		-720	
84	-3900	84	1880
-4500		-4500	
84	600	84	6380
-100		-100	
74	700	74	6480
400		400	
34	300	34	6080
75		75	
31	225	31	6005

(a)   (b)

Figure 7: Heat cascade.

Heating and cooling utilities can be supplied to a process in various ways. An aid for the design and selection of utility systems is the Grand Composite Curve. This tool gives graphical information of the amount of heating and cooling required by the process as well as the temperature at which such utilities can be supplied. The

Grand Composite Curve is constructed from the information obtained from the heat cascade. This is shown in Fig. **8** where the heating duty is supplied in the following way: part of the load is met by means of high pressure steam (HP) and the rest with medium pressure (MP) and low pressure steam. Depending on the cooling needs and the temperatures at which heat is to be eliminated, the cooling load can be removed using cold water, a refrigeration system or by other means such as water preheating for low pressure steam production.

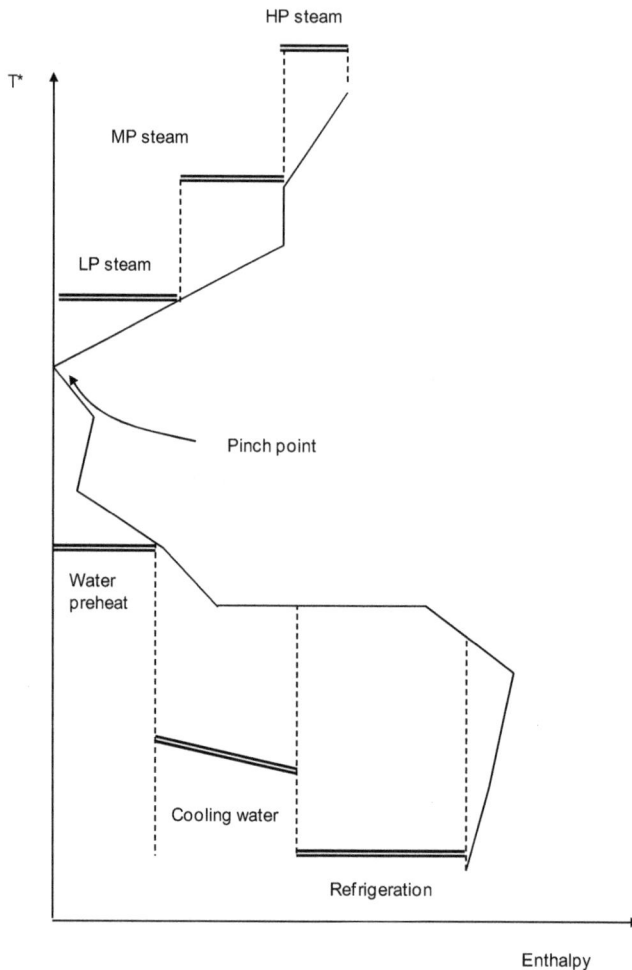

Figure 8: Grand Composite Curve for process utility selection.

When low temperature cooling is required, refrigeration systems are employed. The most common refrigeration systems are: vapor compression systems and absorption cycles. Overall, these refrigeration cycles consist of four steps, which are compression, condensation, expansion and evaporation. In the case of absorption systems, the compression stage is replaced for a system where the coolant is compressed in the liquid phase before it is released at high pressure as vapor. Systems such as NH_3-H_2O and H_2O-$LiBr$ are industrially used. The type of compression mechanism has an implication upon the type of energy source that is used to run it. For instance, a mechanical compression system is run using electrical power, whereas an absorption system uses electrical power for the compression of the liquid and thermal energy to release the vapor at high pressure. Absorption cycles are convenient in cases where there is a surplus of low grade thermal energy. Fig. **9** shows the flow diagram of a simple absorption cooling system.

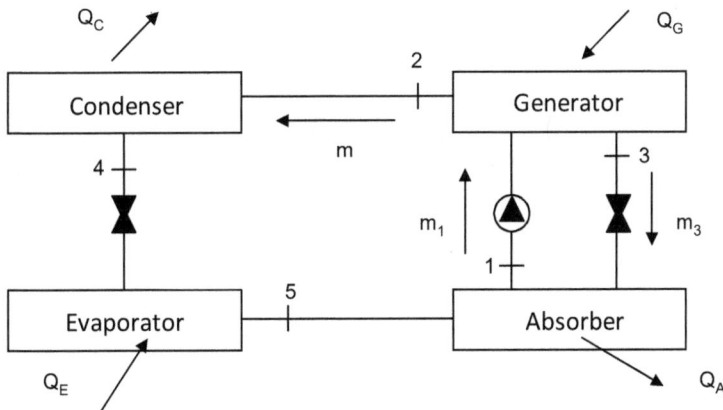

Figure 9: Simple absorption cooling cycle.

The refrigeration load that represents the evaporator duty is given by Q_E. Apart from the electrical energy required to compress the liquid from the absorber, the main energy input to the system is the heat load of the generator (Q_G). The following expressions can be applied to a system where energy losses are negligible:

$$m = \frac{Q_E}{h_5 - h_4}$$

(1)

From a mass balance around the absorber we get:

$$mx_1 = m_3 x_3$$

$$m_1 = m + m_3 \tag{2}$$

Combining these equations we find that:

$$m_3 = \frac{x_1}{x_3 - x_1} m \tag{3}$$

$$m_1 = \frac{x_3}{x_3 - x_1} m \tag{4}$$

From an energy balance around the generator, the absorber and the condenser we respectively get:

$$Q_G = h_3 m_3 + h_2 m - h_1 m_1 \tag{5}$$

$$Q_A = h_5 m + h_3 m_3 - h_1 m_1 \tag{6}$$

$$Q_C = m(h_2 - h_4) \tag{7}$$

Where m is the mass flow rate of the refrigerant, x is the concentration of the refrigerant and h is the specific enthalpy of the refrigerant. For a specified refrigeration load (Q_E), the energy required to run the system is given by Q_G. For industrial applications with temperatures between 4 °C and 12 °C, the H_2O-LiBr systems are used. In situations where the amount of surplus heat surpasses the process needs, the use of steam to run pumping systems would be justified. From an energy balance around a pump we get:

$$W_p = m_p(h - h^*) \tag{8}$$

Where W_p is the power required for pumping, m_p is the steam flow rate, h_1 is the enthalpy of the steam at the inlet of the turbine and h* is the isentropic condition of the turbine inlet steam. In terms of the isentropic efficiency, W_p can be written as:

$$\eta_{iso} = \frac{h_1 - h_2}{h_1 - h^*}$$

(9)

Combining equations (8) and (9) we have:

$$W_p = m_p \, \eta_{iso} \left(h_1 - h^* \right)$$

(10)

5. THERMAL INTEGRATION AND ECONOMICAL ANALYSIS OF THE CO₂ PROCESS

The use of residual energy available in the flue gas stream from where CO_2 is recovered reduces the need for further fuel consumption for the operation of the downstream CO_2 absorption process, thus resulting in energy savings. In this section the thermal integration and an economical analysis is performed in order to determine the feasibility of the process.

From the thermal integration point of view, the process is simple since it consists only of one hot stream (hot flue gas) and one cold stream (bottom of the desorption column). However, since the hot gas stream cannot be used directly to operate the re-boiler of the column, an intermediate heat carrier fluid is needed; so steam can be used. The heat content of the flue gas stream is used to raise medium pressure steam in the following ways: it is used directly to run the re-boiler of the desorption column; it is used to produce power for the operation of pumps and the compression of CO_2; additionally, it can be used for the operation of a LiBr-H_2O absorption system. All this involves the selection of additional pieces of equipment which must be taken into consideration for assessing the economical feasibility of the process.

For the production of medium pressure steam the hot gas is passed through a heat exchanger. The thermal and electrical requirements of the process are functions of the concentration of the MEA solution used as shown in Table **10**. As described above, steam can further be used to run pumps and compressors by means of turbo-machinery. Table **11** shows the direct and indirect energy consumption for various concentrations of MEA.

Table 10: Thermal and electrical requirements of the process

MEA%	Thermal Requirement (kW)	Power Requirement (kW)
10%	36,194.15	1,955.38
15%	32,339.09	1,941.92
20%	30,178.65	1,927.86
25%	28,465.58	1,907.62
30%	27,016.20	1,885.35
35%	25,751.55	1,862.83
40%	24,629.34	1,840.80

The operating conditions of the MEA regeneration column (stripper) demand that the maximum re-boiler temperature be 122 °C due to the thermal sensitivity of MEA. The re-boiler of the regeneration column operates at a temperature of 105.7 °C. This implies that low pressure steam should be used for heating. So, a low steam temperature of 120 °C corresponding to a steam pressure of 2.0 kg/m^2 is used.

Table 11: Steam requirement for direct use and indirect use

MEA%	Direct Use Steam for Heating (kg/s)	Indirect Use Steam for Electrical Needs (kg/s)	Indirect Use Steam for CO$_2$ Compression (kg/s)
10%	13.23	0.70	0.20
15%	11.82	0.70	0.19
20%	11.03	0.69	0.19
25%	10.41	0.69	0.18
30%	9.88	0.68	0.17
35%	9.41	0.67	0.16
40%	9.00	0.66	0.16

From the energy balances it can be seen that 88% of the required energy is in the form of thermal energy and 12% is in the form of power. There exist various options for the thermal integration of the heat recovered from the hot flue gas stream. These include:

a)   For producing 10 bar steam (180 °C); part of this steam should be used to run the turbomachinery and pumps with extraction of low pressure steam (2 bar) should be operated; the rest of the steam is pressure reduced (2 bar) to operate the re-boiler of the regeneration column.

b) For producing 2 bar steam (120 °C) the regeneration column should be operated and surplus steam should be used to run an absorption system.

Option (a) limits the amount of steam produced since the heat content of the flue gases can be extracted only down to 195 °C. A second steam recovery system would be necessary to further recovery of heat. On the other hand, there would only be capacity to run pumps and not the compressor since it operates at a pressure ratio of 20 which cannot be achieved with 10 bar steam. As for option (b), there will be enough low pressure steam to run the re-boiler of the regeneration column and there will be some surplus which can be used to produce cooling through an absorption system. The balance for this option is discussed below.

Typically, the hot flue gases are available at a temperature of 260 °C from where the maximum amount of energy that can be recovered is limited by the pressure at which steam is produced. So, allowing for a minimum temperature approach of 15 °C and targeting for a 2 bar steam, the heat content of the flue gas can then be recovered down to a temperature of 135 °C.

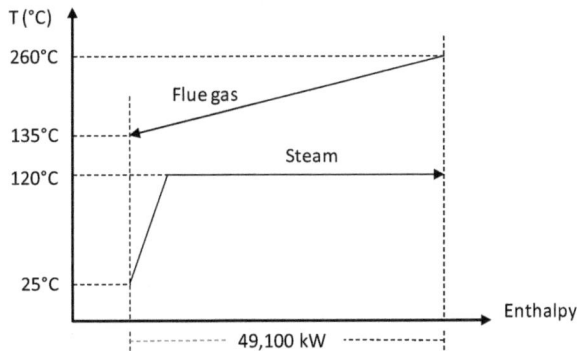

Figure 10: Heat recovery for the production of medium pressure steam.

From the results of the simulation, the energy available for heat recovery between 260 °C and 135 °C is 49,100.0 kW. The effective heating load of steam is 41, 532.34

kW. Fig. **10** shows the temperature *vs.* enthalpy diagram for the heat recovery process. The re-boiler load is 36,194.15 kW; so the whole heat duty can be supplied by heat recovered from the gas steam. The steam surplus that represents an energy content of 5,338.19 kW can be used for running an absorption cooling system and supply part of the cooling duty of the process which is 29,636.74 kW.

Economical Analysis

For the various scenarios, *i.e.*, concentrations of MEA, the following components of the total cost are evaluated:

- Equipment capital cost.

- Operating and maintenance costs.

- Total costs.

The equipment costs were calculated from the results obtained using Aspen Plus. Their costs were updated to the year 2009, and are shown in Table **12**. Fig. **11** shows the variation of the total equipment costs with the molar ratios α (mol CO$_2$/mol MEA).

Table 12: Total equipment costs in USD

Plant Capacity (MW)	Total Equipment Cost in USD for Different MEA Concentrations						
	10%	15%	20%	25%	30%	35%	40%
100	5,952,64 5.95	5,952,64 5.95	5,952,64 5.95	5,952,64 5.95	5,952,64 5.95	5,952,64 5.95	5,952,64 5.95

Fig. **11** indicates that the equipment cost reduces as the molar ratio increases. This happens since, lower flow rates are processed in each part of equipment resulting in smaller sizes. As for the operation, maintenance and labor costs are estimated as 8000 working hours a year. The maintenance costs are estimated as 2.5% and the labor as 1% of the total plant costs, respectively. Fig. **12** shows the variation of the total plant investment with respect to molar ratio. The variation of the total plant investment with respect to MEA concentration is depicted in Fig. **13**. In this case, the results show that between MEA concentrations of 20% and 35% the investment is the highest. Considering a sale price of compressed CO$_2$ as 3.16 USD/kg, Table **13** shows the total sales and the corresponding profit for the recovery and compression of CO$_2$ for commercial purposes. Using this information, the payback time for a plant operating with a 10% MEA concentration and 0.5 molar ratio, is 2.12 years as shown in Table **14**.

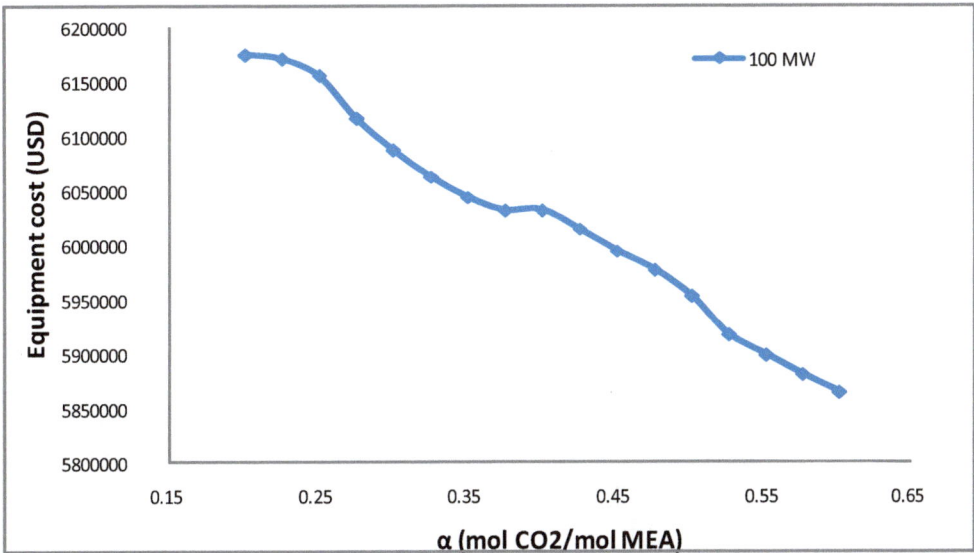

Figure 11: Variation of total equipment costs with respect to molar ratio.

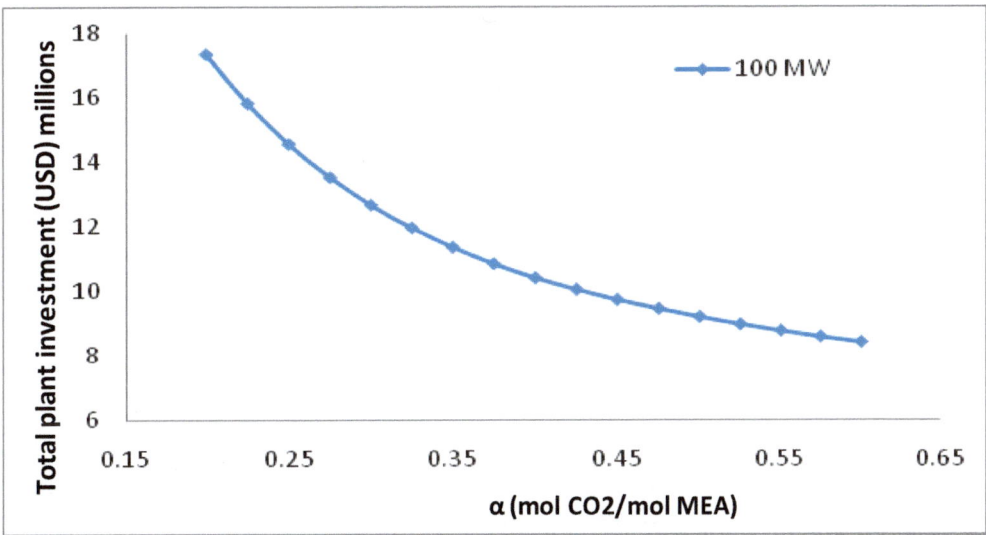

Figure 12: Variation of the total plant investment with respect to α.

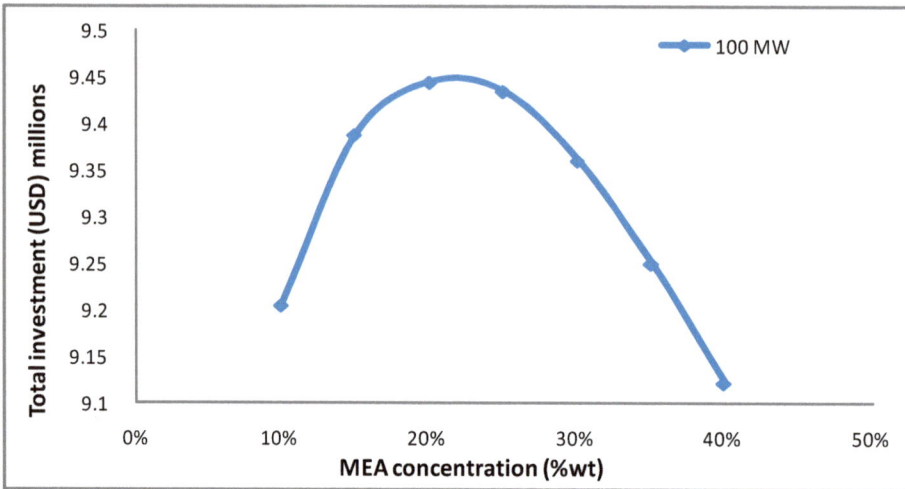

Figure 13: Variation of the total plant investment with respect to MEA concentration.

Table 13: Total profit for the sales of compressed CO_2

%MEA	Total sales (USD)	Operating costs (USD)	Profit (USD)
10%	221,265,345.95	109,933,525.69	111,331,820.25
15%	215,931,184.13	112,140,415.21	103,790,768.92
20%	210,326,591.69	112,831,026.14	97,495,565.55
25%	202,184,659.35	112,711,189.04	89,473,470.31
30%	193,219,391.69	111,821,481.64	81,397,910.05
35%	184,117,266.43	110,506,679.51	73,610,586.92
40%	175,252,178.53	108,974,549.32	66,277,629.22

Table 14: Total investment costs and payback time for a plant operating with 10% MEA concentration and 0.5 molar ratio

Plant Capacity (MW)	Total Investment (USD)	Operating and Maintenance Costs (USD)	Payback Time (Years)
100	126,698,136.2	109,933,525.7	2.12

CONCLUSION

The operation of a CO_2 recovery plant through the absorption of CO_2 in Monoethanolamine (MEA) from a flue gas stream coming from power plants is a well developed technology. The operation of the absorption process is accompanied by the consumption of a large amount of thermal energy. Therefore, energy integration of the upstream power plant with the CO_2 recovery process is vital for the reduction of the operating costs. In this chapter the thermal integration of the flue gas stream has been reviewed and various options for heat recovery are considered. The largest energy consumer in the absorption process is the re-boiler of the MEA regeneration column; it has been shown from the energy balances that this column can be fully driven using the thermal energy recovered from the flue gas stream. From the environmental and economical points of view, the process is feasible with attractive payback times.

ACKNOWLEDGEMENTS

Partial financial support of this work were provided by Consejo Nacional de Ciencia y Tecnología (CONACyT), project: CB-2007-01-82987, EDOMEX-2009-C02-135728 and Chemical Faculty of State of Mexico University in order to use *Aspen Plus*^{TM}. All are greatly appreciated.

CONFLICT OF INTEREST

The authors confirm that this chapter contents have no conflict of interest.

REFERENCES

[1]   Chapel D, Ernst J, Mariz C. *Recovery of CO₂ from flue gases: Commercial trends*, Proceedings of the Canadian Society of Chemical Engineers Annual Meeting, Saskatchewan: Canada **1999**.

[2]   Barchas R, Davis R. The Kerr-McGee/ABB Lummus Crest technology for the recovery of CO₂ from stack gases. *Energ Convers Manage* **1992**, 333-340.

[3]   Sander M T, Mariz C L. The Fluor Daniel 'Econamine FG" process: past experience and present day focus. *Energ Convers Manage* **1992**, 341-348.

[4]   Marion J, Nsakala Y, Bozzuto N, Liljedahl C, Palkes G, Vogel M, Gupta D, Guha J C, Johnson M H, Plasynski S. *Engineering feasibility of CO₂ capture on an existing US coal-fired power plant*. 26th International Conference on Coal Utilization & Fuel Systems, Clearwater, Florida: USA **2001**.

[5]   Alie C, Backham L, Croiset E, Douglas P L. Simulation of CO₂ capture using MEA scrubbing: a flow sheet decomposition method. *Energ Convers Manage* **2005**, 475-487.

[6]   Singh D, Croiset E, Douglas P L, Douglas M A. Techno-economic study of CO₂ capture from an existing coal-fired power plant: MEA scrubbing *vs.* O₂/CO₂ recycle combustion. *Energ Convers Manage* **2003**, 3073-3091.

[7]   Suda T, Fujii M, Yoshida K, Lijima M, Seto T, Mitsuoka S. Development of flue gas carbon dioxide recovery technology. *Energ Convers Manage* **1992**, 317-324.

[8]   Rao A B. *Performance and Cost Models of an Amine-Based System for CO$_2$ Capture and Sequestration.* Report to DOE/NETL, from Center for Energy and Environmental Studies, Carnegie Mellon University, Pittsburgh, PA: USA **2002**.

[9]   Mariz C. Carbon dioxide recovery: large scale design trends. *J Can PetrolTechnol* **1998**, 42–47.

[10]   Trent H, Hoadleyb A, Hoopera B. Process integration analysis of a brown coal-fired power station with CO$_2$ capture and storage and lignite drying. *Energy Procedia 1* **2009**, 3817–3825.

[11]   Kawabata M, Iki N, Kurata O. *Proceedings of the ASME 2010 International Mechanical Engineering Congress & Exposition, IMECE2010,* Vancouver, British Columbia: Canada **2010**; pp. 12-18.

[12]   Amrollahi Z, Ertesvåg I S, Bolland O, Ystad P A M. *Optimized Process Configurations of Post-combustion CO$_2$ Capture for Natural-gas-fired Power Plant — Power Plant Efficiency Analysis, Third International Conference on Applied Energy,* Perugia: Italy **2011**; pp. 629-640.

[13]   Alie C. Simulation of CO$_2$ capture with MEA: Integrating the absorption process and steam cycle of an existing coal-fired power plant. *Energ Convers Manage,* **2004**, 42-57.

[14]   Audus H. *Leading options for the capture of CO$_2$ at power stations, presented at the Fifth International Conference on Greenhouse Gas Control Technologies,* Cairns: Australia **2000**; pp. 13 – 16.

[15]   Riemer P, Audus H, Smith A. *Carbon dioxide capture from power stations, IEA Greenhouse Gas R&D Programme,* Cheltenham: United Kingdom **1993**.

[16]   Gupta M, Coyle I, Thambimuthu K. *CO$_2$ Capture Technologies and Opportunities in Canada, Strawman Document for CO$_2$ capture and Storage (CC&S) Technology Roadmap. CANMET Energy Technology Centre Natural Resources*: Canada **2003**.

[17]   Wong S, Bioletti R. Carbon Dioxide Separation Technologies, *Carbon & Energy Management,* Alberta Research Council Edmonton, Alberta: Canada **2002**.

[18]   *Aspen Plus®* 11.1, *Aspen Technology, User guide, Aspen Physical Property System: Physical Property Methods and Models 11.1.* USA **2001**.

[19]   Kemp I C. *Pinch Analysis and Process Integration.* Oxford, United Kingdom **2007**.

CHAPTER 4

Characterization and Application of Structured Packing for CO_2 Capture

Rosa-Hilda Chavez[1,*], Eva M. de la Rosa[1] and Javier de J. Guadarrama[2]

[1]Instituto Nacional de Investigaciones Nucleares Carretera Mexico-Toluca S/N, La Marquesa, Ocoyoacac, 52750, Mexico and [2]Instituto Tecnológico de Toluca, Av. Instituto Tecnológico de Toluca S/N, Metepec, 52140, Mexico

Abstract: The purpose of this work is to evaluate the minimum energy consumption for solvent regeneration and maximum CO_2 absorption with 600 t/hr flue gas flow simulated by *Aspen Plus™* of CO_2 capture process, using Monoethanolamine (MEA) at 30 weight%. The parameters studied were: 1) energy consumption at reboiler of stripper, 2) absorption separation efficiency, 3) flow ratio (L/G) in order to find the load on turbulence regimen in absorption process, and 4) absorption and stripper column diameters at different flue gas flows. This work contributes structured packing study in separation columns, like: ININ 18, *Sulzer* BX and *Mellapak* 250Y, and the advancement in CO_2 capture technology. Hydrodynamic and mass transfer models were used to evaluate pressure drops and height of mass transfer equivalent unit, per each packing, from experimental data of CO_2 absorption column and predict up to 600 ton/h flue gas flow by *Aspen Plus™*. The results showed that *Sulzer* BX has the highest volumetric mass transfer coefficient values and the lowest height of mass transfer equivalent unit, with $3.76s^{-1}$ and 0.316m, respectively, and the most absorption efficiency of 89.17% in comparison with respect to the other two packings and 120MW reboiler energy.

Keywords: CO_2 capture efficiency, numerical simulation, re-boiler energy consumption at stripper, structured packing, liquid/gas flows ratio.

1. INTRODUCTION

Advanced techniques for capturing and storing of the CO_2 – commonly referred to as CCS – may become an important part of the solution. Within that context the relevant issues are [1]:

Corresponding author Rosa-Hilda Chavez: National Institute of Nuclear Research, Management of Environmental Science, Nuclear Center "Dr. Nabor Carrillo Flores", México-Toluca Road, La Marquesa, Zip Code 52750, Ocoyoacac Estado de Mexico, Mexico; Tel: +52 5553-297200 Ext 12654; E-mail: rosahilda.chavez@inin.gob.mx

- Development of a methodology for optimal design of large scale absorption processes including technical scale-up based on lab- and pilot-scale tests.

- Assessment of existing and emerging solvents with regards to the absorption process performance by means of (flow sheet) simulations.

- Analysis of existing and emerging options/techniques for post-combustion CO_2 removal including regeneration operation and process integration (*e.g.* adsorption, immobilized activator, enzyme catalysis, vacuum stripping for regeneration).

- Assessment and analyses of physiochemical and thermodynamic properties of solvents affecting the efficiency of CO_2 capture (experiments and information available in the open domain).

- Assessment of the need for pre-treatment owing to the impurities in the flue gas.

CO_2 capture from flue gas of post combustion is treated by chemical solvents in packed columns operating in a counter-current mode, with the gas moving in the upward direction and the liquid flowing downward. The packing materials are generally manufactured from ceramics, metal and plastics and divided according to the shape of the gauze packing, grid type, metal sheet and random packings [2-5]. Predictions of pressure drop and mass transfer characteristics are critical for good design. The most important parameters in the research related to the hydrodynamics and mass transfer separation of packed columns are pressure drop, mass transfer effective area, liquid hold-up and mass transfer coefficient [6].

Among all the different techniques for capturing CO_2, absorption with aqueous alkanolamine is recognized as a proper commercial option for capturing CO_2 in gas diluted flows, which contain 10% to 12% of CO_2 volume [7]. The huge amount of flue gas to be treated requires very large size absorbers and high capacity packings. When using chemical solvents, such as amines like the monoethanolamine (MEA), the CO_2 undergoes an equilibrated reaction with the amines in the fast chemical regime. It is thus of high interest to study the influence of liquid physical properties on hydrodynamics and mass transfer in structured packings, as packing geometry and material, liquid physical properties, flow operating conditions, among others, require intensive experiments in limited ranges without being easily able to dissociate the effects of each parameter from one another. In this regime, mass transfer from gas to liquid in an absorbed mass flux is limited by the effective interfacial area, a_e, between the gas and the liquid solvent and requires high mass transfer coefficients, k_G and k_L, in the gas phase and in the liquid phase, respectively [6].

The reaction of CO_2 with alkanolamines has been extensively described in the literature [8-11]. The carbon dioxide capture with MEA aqueous solution consists of contact of the gas stream with amine aqueous solution which reacts with carbon dioxide to form a soluble carbonate salt, by reaction of acid-base neutralization [12]. The simulation of the CO_2 capture process using MEA with *Aspen Plus™*, helps to obtain chemical and physical properties of components, equilibrium properties of ionic and molecular species by the electrolyte-NRTL models; it also facilitates evaluation of different study cases to compare three different structured packings: ININ 18, *Sulzer* BX and *Mellapak* 250Y. This way it is feasible to choose which one fits requirements such as greater CO_2 absorptions as well as lower height of mass transfer unit and less energy consumption for solvent regeneration used.

With all the previous settings in consideration it has been established that the purpose of this work is to evaluate the minimum energy consumption for solvent regeneration and maximum CO_2 absorption with 600 t/hr flue gas flow treated by *Aspen Plus™* of CO_2 capture process, using MEA at 30 weight%.

In the case of primary amines, carbamate is predominantly formed according to the equilibrium reaction (1).

$$CO_2 + 2R\text{-}NH_2 = R\text{-}NH^+_3 + R\text{-}NHCOO^- \text{ where R is } HOCH_2CH_2... \tag{1}$$

2. METHODOLOGY

The main element of a separation column is gas-liquid contactors where mass transfer operation is carried out; the following Table **1** shows the geometric differences of the three studied packing and Table **2** physical properties of gas and liquid flows.

Table 1: Geometric properties of structured packing

Structured Packing	Material	$a(m^2/m^3)$	ε (M^{3}empty Space/M^{3}packed Bed)	θ (°)	FE (a/ε^3)
Sulzer BX	Stainless Steel	498	0.9	45	0.68313
ININ 18	Stainless Steel	418	0.9633	45	0.46834
Mellapak 250Y	Stainless Steel	350	0.86	45	0.56992

Table 2: Physical properties of gas and liquid flows

Flow	Physical Properties	Value
G	ρ_G (kg/m^3)	2.19
	μ_G(kg/ms)	1.76E-05
	D_G(m^2/s)	1.83E-05
L	ρ_L(kg/m^3)	995.02
	μ_G(kg/ms)	4.61E-03
	σ_L(N/m)	6.38E-02
	D_L (m^2/s)	4.01E-04

For the selection of suitable contactors, the following criteria were considered [13]:

1. Pressure drop: This parameter influences both process investment and operational expenses. The second one has a more important effect in operational expenses due to the power consumption to overcome the pressure drop along the packed column.

2. Mass transfer characteristics: Mass transfer influences equipment size and a direct impact on process investment. Mass transfer of the different options is represented by operational values of overall mass transfer coefficient ($k_G a_e$ and $k_L a_e$ [s^{-1}])

3. Efficiency: This parameter is represented as the specific contact area (a [m^2/m^3]), height of a theoretical plate (HTU_{OG} and HTU_{OL}) and the

number of transfer units (*NTU*). It influences process investment and the purity of the clean flue gas leaving the contactor.

4. Reliability: It is based on proven technology, applications in industry, and size of operation.

The methodology was divided in two sections:

1) The hydrodynamic analysis, performed by exploring different regions of operation: preload, loading and flooding regimens, in order to find *L/G* flow ratio per each packing in order to ensure loading or turbulent regimen and optimum mass transfer operation.

The hydrodynamics of each packing was obtained by determining the pressure drop over packed bed height, *ΔP/Z*, due to the passage of gas through the packed bed, either dry (zero liquid flow) or with liquid irrigated flow [14]. Pressure drop in packed columns is always lower than the wet pressure drop measured, because the liquid flowing through the column changes the bed structure due to liquid hold-up [15].

$$\frac{\Delta P_{irr}}{\rho_L gZ} = \frac{\Delta P_{dry}}{\rho_L gZ} \, xAxB, \text{ where } \dots \tag{2}$$

$$A = \frac{\left\{ 1 - \varepsilon \left[1 - \frac{h_o}{\varepsilon} \left[1 + 20 \left(\frac{\Delta P_{irr}}{\rho_L gZ} \right)^2 \right] \right] \right\}^{\frac{2+c}{3}}}{1 - \varepsilon}, \text{ and } \dots \tag{3}$$

$$B = \left[1 - \frac{h_o}{\varepsilon} \left[1 + 20 \left(\frac{\Delta P_{irr}}{\rho_L gZ} \right)^2 \right] \right]^{-4.65} \dots \tag{4}$$

The gas flow percentage with respect to flooding region was determined by the following equation:

$$\%G_{op} = \frac{G_{op}}{G_{flood}} x100 \dots \tag{5}$$

The liquid hold-up in a wetted packed column is formed of static and the dynamic hold-up. The static hold-up $h_{L,stat}$ is the liquid that remains in the packing after irrigation is stopped by capillary and adhesion forces [16]. It is the liquid prevailing in the pores and gaps of the packing. The static liquid hold-up is a function of the physical properties of the liquid: density ρ_L, dynamic viscosity μ_L, and surface tension σ; the characteristic packing properties: void fraction ε, and specific packing surface a; the liquid load u_L, and the acceleration of gravity g [16]. On the contrary, dynamic hold-up is the liquid participating in the liquid flow and raising the liquid velocity. An increase in the dynamic hold-up normally results in better mass transfer [16]. The liquid hold-up associates hydrodynamic parameter for gas/liquid flow in packed beds and the pressure drop and the fluid effective velocity inside the packing.

Wetted surface area a_e is essential for two-phase flow in irrigated packing and in experimental mass transfer. It is closely linked to effective interfacial area because the wetted area can be effective for mass transfer [17]. The difference between the wetted surface and the effective interfacial area lies in that the wetted surface area incorporates liquid surface area in dead zones and the effective interfacial area includes surfaces of drops and jets [15, 17].

2) Mass transfer model was developed to calculate and analyze height of mass transfer unit (*HTU*) on the gas and liquid phases. The Double Film theory correlates height of global mass transfer unit HTU_{OG} and HTU_{OL}, with height of gas mass transfer unit HTU_G and liquid mass transfer unit HTU_L for a system [17, 18].

The effective gravity is defined as:

$$g_{eff} = 9.81 \left[\left(\frac{\rho_L - \rho_G}{\rho_L} \right) \left(1 - \frac{\Delta P / \Delta Z}{(\Delta P / \Delta Z)_{flood}} \right) \right] \dots$$

(6)

The liquid hold-up in a wetted packed column is defined as:

$$h_L = \left(4 \frac{F_t}{S} \right)^{2/3} \left(\frac{3\mu_L (U_L)}{\rho_L (Sin\theta) g_{eff} (\varepsilon)} \right)^{1/3} \dots$$

(7)

The corrugated side of packing is:

$$S = \frac{4\varepsilon}{a_p} \quad \ldots \tag{8}$$

The gas and liquid effective velocities are defined as:

$$U_{Leff} = \frac{U_L}{\varepsilon(Sin\theta)h_t} \quad \ldots \tag{9}$$

$$U_{Geff} = \frac{U_G}{\varepsilon(Sin\theta)(1-h_t)} \quad \ldots \tag{10}$$

These calculations show that the flow dynamics must be known in details in particular for determining mass transfer characteristics. Indeed, the liquid velocity at the interface is further used for the gas and liquid side mass transfer coefficient, k_G, k_L, determination by Higbie theory [17-20].

$$\left(\frac{K_G S}{D_G}\right) = 0.054 \left[\frac{\left(U_{L,eff} + U_{G,eff}\right)\rho_G S}{\mu_G}\right]^{0.8} \left(\frac{\mu_G}{D_G \rho_G}\right)^{0.33} \quad \ldots \tag{11}$$

$$k_L = 2\sqrt{\frac{D_L\left(U_{Leff}\right)}{\pi(S)0.9}} \quad \ldots \tag{12}$$

$$\frac{a_e}{a} = F_{SE}\left[\frac{29.12\left(We_L\, Fr_L\right)^{0.15} s^{0.359}}{Re_L^{0.2}\, \varepsilon^{0.6}\left(1-Cos\gamma\right)\left(Sen\theta\right)^{0.3}}\right] \quad \ldots \tag{13}$$

On the bases of conventional definitions of transfer units, the height of a gas phase transfer unit is:

$$HTU_G = \frac{U_G}{K_G a_e \rho_G} \quad \ldots \tag{14}$$

And the height of a liquid phase transfer unit is:

$$HTU_L = \frac{U_L}{K_L a_e \rho_L} \ldots \tag{15}$$

The application of the two-film model to mass transfer is based on the number of transfer global units, *NTU*, of both gas and liquid resistances, and it involves the efficiency in terms of the height of a transfer global unit *HTU* [21-23].

By the gas phase side: $Z = HTU_{OG} * NTU_{OG} \ldots$ (16)

And by the liquid phase side: $Z = HTU_{OL} * NTU_{OL} \ldots$ (17)

The application of the two-film model is frequently used to relate the height of the transfer global unit (HTU_{OG} or HTU_{OL}) with the height of the gas HTU_G and liquid HTU_L transfer units to the absorption.

By the gas-side: $HTU_{OG} = HTU_G + \lambda HTU_L \ldots$ (18)

And by the liquid-side: $HTU_{OL} = HTU_L + \dfrac{1}{\lambda} HTU_G \ldots$ (19)

If the gas is highly soluble in liquid, the Henry's constant will be small. In this case the liquid-side resistance is negligible. If the gas is relatively insoluble (Henry's constant will be large) the gas-side resistance becomes negligible in comparison with the liquid-side resistance. The relative magnitude of the individual resistance evidently depends on gas solubility, as represented by the Henry's law constant. This explains the common statements that "the liquid side resistance has the control" in the absorption of a relatively insoluble gas, and the "gas-side resistance has the control" when a relatively soluble gas is absorbed (or stripped) [24].

The term λ is the ratio of slopes, equilibrium line to operating line and it is known as the removed factor:

$$\lambda = m \frac{U_G}{U_L}, \text{ where } m = \frac{H}{P_T} \ldots \tag{20}$$

The inverse of λ is known as the absorption factor A.

The global mass transfer coefficient by gas-side is: $K_G = \dfrac{1}{\dfrac{1}{k_G} + \dfrac{H}{k_L}}$ ...   (21)

The global mass transfer coefficient by liquid-side is: $K_L = \dfrac{1}{\dfrac{1}{k_L} + \dfrac{1}{Hk_G}}$ ...   (22)

Gas resistance is: $\% \,\mathrm{Re}\,sis\,tan\,ce_G = \dfrac{\dfrac{1}{k_G}}{\dfrac{1}{K_G}}(100)$ ...   (23)

Liquid resistance is $\% \,\mathrm{Re}\,sis\,tan\,ce_L = \dfrac{\dfrac{1}{k_l}}{\dfrac{1}{K_l}}(100)$ ...   (24)

3. CONCEPTUAL PROCESS DESIGN OF CO_2 CAPTURE PROCESS

The following input data are required to simulate the absorber and stripper system:

1) Solvent type and composition (density and heat capacity);

2) Gas and liquid inlet temperatures;

3) Inlet and outlet CO_2 concentrations of the gas stream;

4) Pure components properties and mixture properties (physical properties);

5) Vapor – Liquid equilibrium of CO_2 of the solvent at absorption and regeneration;

6) Vapor pressure of the solvent;

7) Reaction and absorption enthalpy. Mass transfer parameters: k_G, k_L and a_e with the selected contactor type and the solvent. CO_2 is converted into ionic species. They are non-volatile and remain in the liquid phase until CO_2 is released in the stripper by adding heat. Before the length of the absorber can be calculated from the input parameters, the mass transfer parameters (k_L, k_G and a_e) need to be known. These parameters have been estimated from empirical correlations as available in literature (Equations 11, 12 and 13).

When these mass transfer parameters were calculated, the absorber performance can be simulated.

The following data are calculated by the model:

1) Required absorber dimension;

2) Temperature profile in the absorber;

3) Concentration profile in the absorber;

4) Profile of the speciation in the liquid phase.

The quality of a process model such as mass transfer aspects, thermodynamics and kinetics should always be validated with experimental data.

4. PILOT PLANT SIMULATIONS

The gas flow G_1 enters at the bottom of the absorption column, while the liquid flow L_1 entering at the top. The rich liquid amine stream L_2 enters a heat exchanger (CALENT) to raise its temperature and then enters desorption column where it carries out the regeneration of the amine. The regenerated liquid flow is mixed with pure solution of MEA (30 wt%), L_5, in order to comply with the mass balance of the whole process. Due to stream L_6, it enters a heat exchanger to decrease its temperature (ENFRIA) and then re-circulates (L_7) in the absorption column. The stability of the stream L_5 is the clue to find out the iteration number in the simulator (see Fig. **1**). CALENT and ENFRIA could be the same heat exchanger as energy saver equipment.

The *Aspen Plus*™ simulator [25] requires input data from gas stream such as temperature, pressure, flow, and composition, as well as diameter, the number of stages, pressure, and some parameters for each of the three packings studied for the two columns (absorption and desorption) and MEA regeneration re-boiler energy.

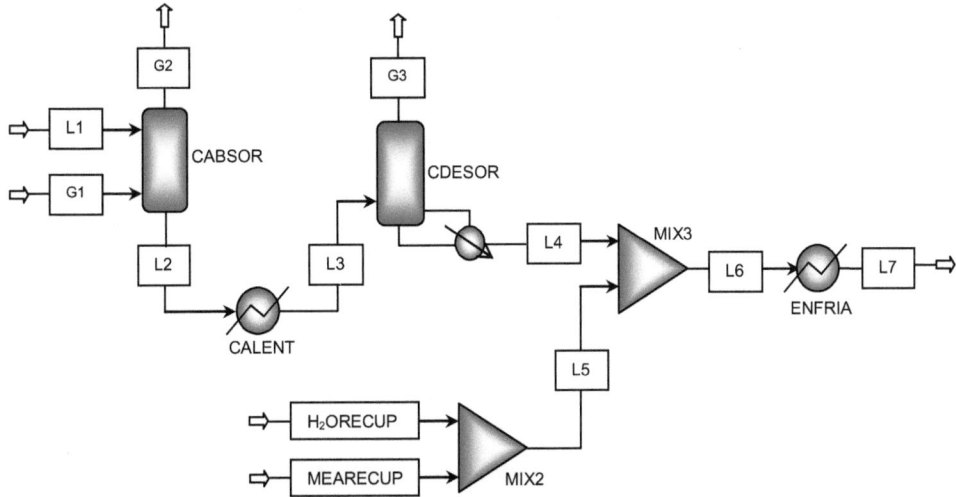

Figure 1: Flow chart of CO$_2$ capture process.

To reduce the complexity of simulation in *Aspen Plus*™, the following considerations were done: i) the washing column to filter out particles and the stage of desulfurization of flue gases to remove sulfur dioxide, were not included in the simulation process, ii) The inlet gas stream was considered free from all contaminants and consisted of H$_2$O, CO$_2$, O$_2$ and N$_2$; NO$_x$ and SO$_2$ were not considered, iii) Corrosion was not taken into account, iv) Storage and transport of CO$_2$ were not included.

CO$_2$ absorption column and CO$_2$ capture process efficiencies were calculated as follows:

$$Abs.\ Eff. = \frac{x_{G1}^{co_2} - x_{G2}^{co_2}}{x_{G1}^{co_2}}(100)$$

$$\dots \qquad (25)$$

$$Cap.\ Eff. = \frac{x_{G3}^{co_2} - x_{G1}^{co_2}}{x_{G3}^{co_2}}(100)$$

$$\dots \qquad (26)$$

RESULTS AND DISCUSSIONS

A large number of experiments have been carried out with aqueous MEA in the pilot plant. In this study, experiments in a CO_2 capture pilot plant which was operated at typical post-combustion flue gas conditions were compared quantitatively with experimental data derived from a continuous absorber pilot plant. The results of these experiments have been compared to the outcome of the simulations of a rate-based absorption model.

Pressure drop, liquid hold-up and effective mass transfer area have been successfully determined in a small pilot plant absorber of 0.3 m ID, 3.5 m packing height as function of gas and liquid flow rate. The hydraulic variable and effective mass transfer area parameter dependencies were in reasonable agreement with correlations found in literature.

Mass transfer, reaction kinetics and thermodynamics are included in the simulation model. With this model it was possible to predict the size of an absorber within an accuracy of 10% [26]. Also the CO_2 flux, chemical enhancement and liquid speciation were calculated through the model.

The mass transfer model with complex, reversible, multiple chemical reactions included, describes the absorption of CO_2 in an aqueous solution of an amine (30 wt%). The absorption rate of CO_2 is calculated in combination with the concentration profiles near the gas-liquid interface for all reactants and products. In the model, kinetic data of the several chemical reactions are incorporated and the thermodynamic model used is based on ideal activities [21, 26].

Table **3** shows pressure drops measurement along packing height of 3.5 m and 0.3 ID, and liquid and gas flow rates, and other hydrodynamic and mass transfer parameters.

Statements like "resistance by the liquid side is controlled for a relatively insoluble gas", and then the liquid side is used for mass transfer evaluation. In case of "resistance of gas side is controlled for a relatively soluble gas", the gas side is used for mass transfer evaluation [27]. The mass transfer resistance was presented by gas side and was observed to be larger than liquid side, so the global mass transfer height was taken on gas side.

Table 3: Hydrodynamic and mass transfer results by a small pilot plant absorber of 0.3 m ID, 3.5 m packing height, as function of gas and liquid flow rate using Bravo model [18]

Flows (kg/h)		*G*	*L*	*G*	*L*	*G*	*L*
		115.00	**345.00**	**115.00**	**345.00**	**115.00**	**345.00**
		Sulzer BX		ININ 18		*Mellapak* 250Y	
$\Delta P_{op}/Z$	Nm^{-2}/m	472.93		791.64		422.25	
g_{eff}	m/s^2	7.07		5.39		7.37	
h_L		0.1940		0.1765		0.1691	
S	m	0.0072		0.0092		0.0097	
We_L		0.0069		0.0088		0.0093	
Fr_L		0.0009		0.0007		0.0006	
Re_L		12.25		15.59		16.46	
F_t		0.9918		0.9894		1.0758	
$Cos\gamma$		0.4402		0.4402		0.4402	
$U_{L,eff}$	m/s	0.0636		0.0653		0.0772	
$U_{G,eff}$	m/s	2.32		2.12		2.38	
U_r	m/s	2.38		2.18		2.46	
M	H/P	1.26		1.26		1.26	
Λ		0.66		0.66		0.66	
HTU_{OG}	m	0.3167		0.4245		0.4297	
HTU_{OL}	m	0.4777		0.6405		0.6483	
$k_G a_e$	s^{-1}	3.763		2.8066		2.7730	
$k_L a_e$	s^{-1}	5.5211		4.159		4.0024	
Gas resistance	%	54.19		54.44		53.79	
Liquid resistance	%	45.80		45.55		46.21	

As Sherwood [27] indicates when a relatively soluble gas is absorbed (% gas resistance is bigger than % liquid resistance), the gas side is controlling, and in this case HTU_{OG} is taken into account.

The HTU_{OG} value to be used in the simulator was the average of the three global mass transfer heights of each package, as 0.4 m, in order to be evaluated under the same overall height but using geometric values as factor packing, geometrical area and porosity per each package.

Tables **4** and **5** show feed data for simulation evaluation under 600 t/hr of G$_1$ and 540 t/hr of MEA plus 1260 t/hr of H$_2$O, using *Sulzer* BX, ININ 18 and *Mellapak* 250Y with L/G of 3 in order to be in a load regimen.

Table 4: Feed data for simulation evaluation under 600 t/hr of G1 and 540 t/hr of MEA plus 1260 t/hr of H$_2$O

Stream	Stream Properties	Value
G$_1$	Temperature (°C)	55
	Pressure (bar)	1
	Mass flow (t/h)	600
	H$_2$O (mass fraction)	17.5
	CO$_2$ (mass fraction)	9.8
	N$_2$(mass fraction)	66.0
	O$_2$ (mass fraction)	6.7
H$_2$O	Temperature (°C)	35
	Pressure (bar)	1
	Mass flow (t/h)	1260
MEA	Temperature (°C)	35
	Pressure (bar)	1
	Mass flow (t/h)	540
H$_2$ORECUP	Temperature (°C)	35
	Pressure (bar)	1
	Mass flow (t/h)	23.3
MEARECUP	Temperature (°C)	35
	Pressure (bar)	1
	Mass flow (t/h)	10

Table 5: Feed data for simulation evaluation under 600 t/hr of G$_1$ and 540 t/hr of MEA plus 1260 t/hr of H$_2$O

Blok Name	Blok Properties	Value		
MIX1, MIX2, MIX3	Type	MIXER		
	Pressure (bar)	1		
CALENT	Type	HEATER		
	Temperature (°C)	90		
	Pressure (bar)	1		
ENFRIA	Type	HEATER		
	Temperature (°C)	35		
	Pressure (bar)	1		
		Sulzer BX	ININ 18	*Mellapak* 250Y
CABSOR	Type	RATEFRAC	RATEFRAC	RATEFRAC
	Pressure (bar)	1	1	1
	Condensate	No	No	No
	Reboiler	No	No	No
	Column diameter (m)	1.1	1.1	1.1
	Stage number	8	7	7
CDESOR	Type	RATEFRAC	RATEFRAC	RATEFRAC
	Pressure (bar)	1	1	1
	Condensate	Not	Not	Not
	Reboiler	Yes	Yes	Yes
	Column diameter(m)	1.1	1.1	1.1
	Stage number	8	7	7

ININ18 packing presented better hydrodynamic behavior than the other two because ININ18 operated at 80% from flooding point and *Mellapak* 250Y and *Sulzer* BX at 60% from flooding point at the same $L/G = 3$ flow ratio for three structured packings (see Fig. **2**). ININ18 packing showed higher pressure drop than the other two structured packings. This one reached the flood with lower liquid and gas flow values. Table **5** shows the hydrodynamic and mass transfer parameters with the same $L/G = 3$, at loading regimen. *Sulzer* BX is the lowest height of mass transfer unit and the highest value on the volumetric mass transfer coefficients, due to its geometric characteristics.

The following values were obtained:

- Liquid side mass transfer coefficient k_L. $a_e = 5.5$ s^{-1} for *Sulzer* BX, 4.15 s^{-1} for ININ 18 and 4.0 s^{-1} for *Mellapak* 250Y; these values are based on an average of the k_L and a_e calculated according to the Equations 11, 12 and 13.

- Gas side mass transfer coefficient k_G. $a_e = 3.7$ s^{-1} for *Sulzer* BX, 2.8 s^{-1} for ININ 18 and 2.7 s^{-1} for *Mellapak* 250Y; these values are based on an average of the k_G and a_e calculated according to the Equations 11, 12 and 13.

The lowest mass transfer height of *Sulzer* BX was 0.32 times less than ININ18 and 0.38 than *Mellapak* 250Y, and the highest effective area of *Sulzer* BX was 0.16 times greater than ININ18 and 0.23 than *Mellapak* 250Y.

Also, *Sulzer* BX shows the highest value on the volumetric mass transfer coefficients of 3.76s^{-1}. *Sulzer* BX packing was the most mass transfer efficient due to the lowest HTU_{OG} value than the other two structured packings, due to the highest geometric area.

The energy requirement for 600 t/h is 120 MW (see Fig. **3**). Simulations were made at different reboiler energies and absorption column and CO$_2$ capture efficiencies were evaluated with respect to CO$_2$ concentration at gas streams by Equations 18 and 19. Three structured packings have the same graphical tendency; however the *Sulzer* BX presents the highest values in all cases. The power requirement in the reboiler is linked to power generation plant, so despite higher power greater capture efficiency, not much energy can be used for the CO$_2$ treatment. It was considered a reboiler energy requirement of 120MW as a more pronounced slope value of CO$_2$ capture efficiency for the three structured packings (see Fig. **3**).

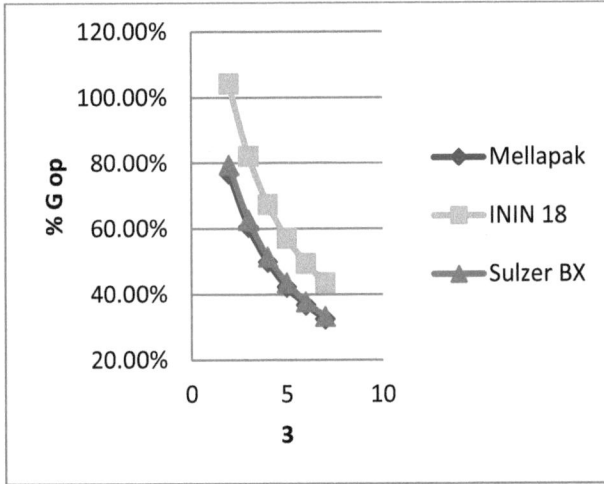

Figure 2: The operated gas flow percentage *versus* L/G flow ratio.

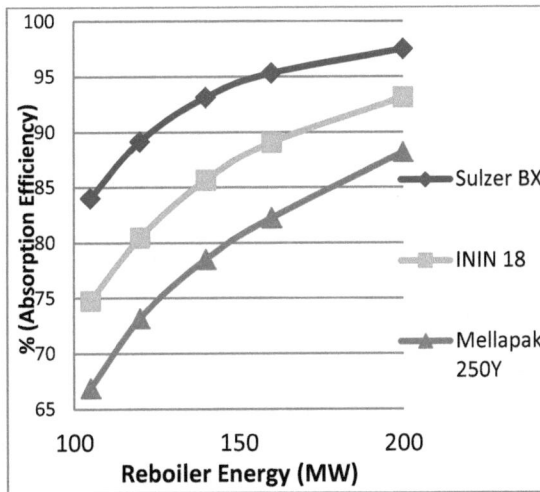

Figure 3: Study cases of energy requirement in the reboiler.

Table **6** shows *Sulzer* BX requires less recovery solution $L_5 = 50.52 \, t/h$, releases less CO_2 into the atmosphere $x_{G_2}^{CO_2}$ equal to 0.0106 and provides higher CO_2 concentration at G_3 stream of $x_{G_3}^{CO_2}$ equal to 0.2839.

Table **7** shows the simulations results for three structured packings under the same conditions. *Sulzer* BX packing was the most efficient in CO_2 whole capture with MEA and it showed greater efficiency in the absorption column, although it requires a number bigger for mass transfer stages. The mass transfer height was however lower in both columns but with highest CO_2 absorption efficiency and CO_2 capture efficiency.

Table 6: Flows and CO_2 mass concentration at different streams

	Sulzer BX	ININ18	*Mellapak* 250Y
$G_1\,(t/h)$	600.00	600.00	600.00
$x_{G_1}^{CO_2}$	0.098	0.098	0.098
$G_2\,(t/h)$	471.150	475.038	477.888
$x_{G_2}^{CO_2}$	0.0106	0.0191	0.0263
$G_3\,(t/h)$	179.371	178.902	178.349
$x_{G_3}^{CO_2}$ (on wet basis)	0.2839	0.2606	0.2412
$L_1\,(t/h)$	1800.00	1800.00	1800.00
$L_2\,(t/h)$	1928.850	1924.962	1922.113
$L_3\,(t/h)$	1928.850	1924.962	1922.113
$L_4\,(t/h)$	1749.480	1746.061	1743.764
$L_5\,(t/h)$	50.520	53.939	56.236
$L_6\,(t/h)$	1800.00	1800.00	1800.00

Figs. **4** and **5** show the exponential increase of the column diameter of absorber and desorber columns. The total flow of 2640 t/h (600 t/h of gas and liquid 2040 t/h) gave a column diameter of 1.1m. The average value of rich load was 0.43, which meant good CO_2 capture; and the average of lean load was 0.30, meant a good regenerated MEA.

The energy requirements of 120 MW for the CO_2 capture for a Power plant of 616MW are:

$$\therefore \quad \frac{120\,MW}{616\,MW}*100 = 19.48\%$$

All structured packings of CO_2 capture had similar values of capture efficiency values, enriched charge (CA) and impoverished load (CD). The *Sulzer* BX packing required a further stage in the absorption and desorption column which meant greater column height (0.4 m more height than ININ 18 and *Mellapak* 250Y packings), but this one had higher absorption efficiency of 89.17% (1.11 times greater than ININ 18 and 1.22 times greater than *Mellapak* 250Y).

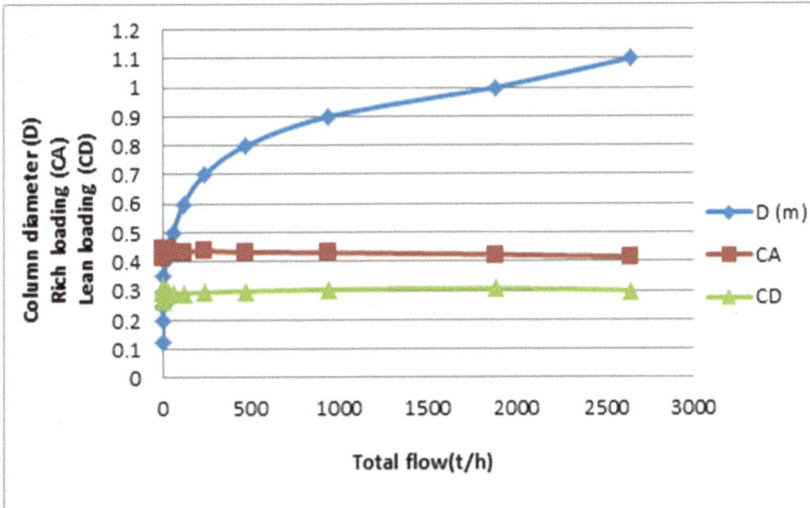

Figure 4: Column diameter, rich loading and lean loading for absorption and desorber columns, respectively, with respect to the total flow.

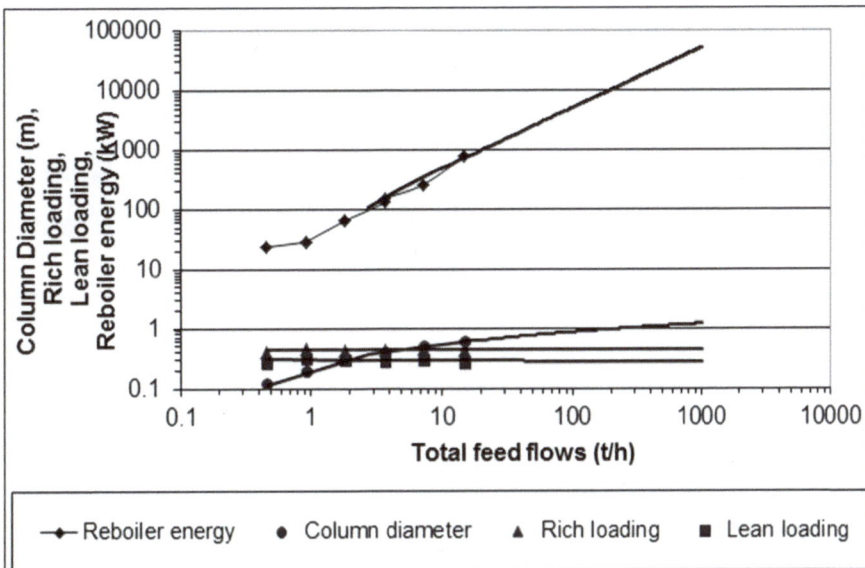

Figure 5: Column diameter, rich loading and lean loading and reboiler energy for absorption and desorber columns, respectively, with respect to the total flow.

Table 7: Simulation with *Aspen Plus*[TM]

Packing	FT (t/h)	L (t/h)	G (t/h)	d (m)	ER (MW)	E	CA	CD	EC_{bs} %	EA %
Sulzer BX	2400	1800	600	1.1	120	8	0.395	0.277	90.04	89.17
ININ 18	2400	1800	600	1.1	120	7	0.389	0.282	90.01	80.47
Mellapak 250Y	2400	1800	600	1.1	120	7	0.381	0.283	89.99	73.17

CONCLUSION

-   ININ18 packing showed higher pressure drop than the other two structured packings. This one reached the flood with lower liquid and gas flow values.

-   *Sulzer* BX packing was the most mass transfer efficient due to the lowest HTU_{OG} value than the other two structured packings due to the highest geometric area.

-   *Sulzer* BX packing was the most efficient in CO_2 whole capture with MEA and it showed greater efficiency in the absorption column, although it requires an extra mass transfer stage. It was lower mass transfer height in both absorption and stripper columns with the highest CO_2 absorption efficiency and CO_2 capture efficiency.

-   The minimum consumption for solvent regeneration was 120MW at reboiler duty in order to carry out the regeneration of the MEA.

ACKNOWLEDGEMENTS

Partial financial support of this work was provided by Consejo Nacional de Ciencia y Tecnología (CONACyT), project: CB-2007-01-82987, EDOMEX-2009-C02-135728 and Chemical Faculty of State of Mexico University in order to use *Aspen Plus*[TM]. All are greatly appreciated.

CONFLICT OF INTEREST

The authors confirm that this chapter contents have no conflict of interest.

NOMENCLATURE

a	Geometric area of structure packing	m^2/m^3
a_e	Effective area as wetted surface area	m^2/m^3
CA	Rich load at absorption column	$MolCO_2$/mol MEA
CD	Lean load at stripper column	$MolCO_2$/mol MEA
$Cos\,\gamma$	Contact angle	
d	Column diameter	M
D	Diffusivity	m^2/s
d_p	Equivalente diameter	M
E	Absorption and stripper stages	
EA	Absorption efficiency	%
EC	Capture efficiency	%
ER	Reboiler energy consumption	MW
FE	Packing factor	M
Fr	Froude number	
FT	Total flow	Tons/h
g	Acceleration of gravity constant	m/s^2
G	Gas flow	[kg/h] or [t/h]
$\%G$	Percentage of gas flow from flooding gas flow	
h_L	Total hold up	
H	Henry´s constant	Atm
HTU	Global mass transfer unit height	M
k	Coeficiente local de transferencia de masa	m/s
K	Mass transfer coefficient	m/s
L	Liquid flow	[kg/h] or [t/h]
L/G	Flow ratio	[(kg/h)/(kg/h)]
P	Pressure	[atm] [bar]
PM	Molecular weight	g/gmol
P_T	Total pressure	atm

Re	Reynolds number	
S	Corrugated side	M
T	Temperature	°C
T	Tons	1000 kg
U	Flow velocity	m/s
We	Weber number	
X	Composition, fraction mol	
Z	Total packed height	M

ε	Void fraction	m^3/m^3
θ	Corrugated angle	
ρ	Density	kg/m^3
σ	Superficial tension	N/m
μ	Kinetic viscosity	m^2/s
λ	Ratio of equilibrium line over operating line	
ΔP	Pressure drop	N/m^2
Subscripts		
dry	Dry packing	
Eff	Effective	
G	Gas	
flood	Flooding	
irr	Irrigated	
L	Liquid	
op	Operation	
R	relative	
Stat	Static	
1,., 7	Stream or flow	

REFERENCES

[1]   Stangeland A., A model for the CO_2 capture potential, *International Journal of Greenhouse Gas Control* 1, **2007**, 418-429.
[2]   Billet R., *Packed Towers*, VCH (eds). Weinheim, Germany, **1995**
[3]   Tobiesen F.A, and Svendsen H.F., *AICHE J.* **2007**, 53(4), 846-854.

[4]   Tsai R.E., Schulthesiss P., Kettner A., Lewis J.C., Seibert A.F, Elridge R.B. and Rochelle G.T., *Ind. Eng. Chem. Res.* **2008**, 47, 1253-1261.

[5]   Alix P. and Raynal L, *Chem. Eng. Res and Design* **2008**, 86, 585-593.

[6]   Raynal L., Ben-Rayana F. and Royon-Lebeaud A., Use of CFD for CO_2 absorbers optimum design: from local scale to large industrial scale, *Energy Procedia* 1, **2009**, 917-924.

[7]   Danckwerts P.V. and McNeil K.M.,The absorption of Carbon Dioxide into aqueous amine solutions and the effects of catalysis, Trans. *Inst. Chemical Engineering*, **1967**,45, T32.

[8]   Danckwerts P.V., The reaction of CO_2 with ethanolamines, *Chem. Eng. Sci*, **1979**, 34, 443-446.

[9]   Blauwhoff P., Versteeg G., Van Swaaij W., A study on the reaction between CO_2 and Alkanolamines in aqueous solutions, *Chem. Eng. Sci*, 1984, 39, 207-225.

[10]   Versteeg G.F., Duijk van L.A.J and Swaaij van W.P.M., On the kinetics between CO_2 and Alkanolamines both in aqueous and non-aqueous solutions, *An overview, Chem. Eng. Comm.* **1996**, 144, 113-158.

[11]   Barth, D.,Tondre, C., Delppuech, J.-J., Kinetics and mechanisms of the reactions of carbon dioxide with alkanolamines: a discussion concerning the cases of MDEA and DEA, *Chem. Eng. Sci.* (39) **1984**, 1753-1757.

[12]   Asrarita G., 1964, The influence of carbanation ratio and total amine concentration on carbon dioxide absorbtion in aqueous monoethanolamine solutions, *Chem. Eng. Sci*, **1964**, 19, 95-103.

[13]   Sanchez E.S. and Goetheer E.L.V., *Energy Procedia* 4, **2011**, 868-875.

[14]   Stichlmair J., Bravo J.L., and Fair R.J., General model for prediction of pressure drop and capacity of countercurrent gas/liquid packed columns. *Gas Sep Purif*, **1989**, 3, pp. 19-28.

[15]   Zakeri A., Einbu A., Wiig P.O., Oi L.E. and Svendsen H.F., Experimental investigation of pressure drop, liquid hold up and mass transfer parameters in 0.5m diameter absorber column, *Energy Procedia* 4, **2011**, 606-613.

[16]   *Ullmann's encyclopedia of industrial chemistry*.; Eds.: John Wiley & Sons Inc. Wiley-VCH & Co. KGaA,Weinheim, **2005**.

[17]   Shi, M. G.; Mersmann, A. Effective Interfacial Area in Packed Columns. *Ger. Chem. Eng.* **1985**, 8, 87.

[18]   Bravo J.L., Rocha J.A. and Fair J.R., Mass transfer in gauze packings, *Hydrocarb Process*, **1985**, 64(1), 91-95.

[19]   Onda K., Takeuchi H. and Okumoto Y., Mass Transfer Coefficients between Gas and Liquid Phases in Packed Columns, *J. Chem. Eng.* Japan 1, **1968**, 1, 56-62.

[20]   Spiegel L. and Meier W., *Correlations of the performance characteristics of various Mellapak types (capacity, pressure drop and efficiency), Distillation & Absorption* Symposium Series No.104, **1987**, 1.

[21]   Huttenhuis P.J.G., Van Elk E.P. and Versteeg G.F, Mass transfer in a small scale post-combustion flue gas absorber, experiment and modeling, *Energy Procedia* 1, **2009**, 1131-1138.

[22]   Henley, Ernerst J. y Seader, J.D., 1981. *Equilibrium-Stage Separation Operation in Chemical Engineering.* Eds.: John Wiley & Sons, New York. **1981**, 638.

[23]   Hines A. and Maddox R., *Mass Transfer*, Prentice Hall, 1985

[24]   Welty, J.R., Wicks, C.E., Wilson, R.E. & Rorrer, G.L., *Fundamentals of Momentum, Heat and Mass Transfer*, 5ᵗʰ ed.;, Eds.: Wiley and Sons, **2008**.

[25] *Aspen Plus, Aspen Plus user guide, Aspen technology limited*, Cambridge, Massachusetts, USA 2003.

[26] Leites I.L., Sama D. A. and Lior, The theory and practice of energy saving in the chemical industry: some method for reducing thermodynamic irreversibility in chemical technology processes, *Energy,* **2003**, 28, 55 – 97.

[27] Sherwood T.K., Pigford C.L. and Wike C.R., *Mass transfer*, International Student Edition, Eds.: Mc. Graw Hill, Japan, **1975**.

CHAPTER 5

Calcium Looping Technology for CO$_2$ Capture

Pilar Lisbona, Yolanda Lara, Ana Martínez and Luis M. Romeo[*]

CIRCE Institute - University of Zaragoza, C/Mariano Esquillor Gomez, 15, 50018-Zaragoza, Spain

Abstract: This technology makes use of the idea that lime may be reused in a cyclic process to remove CO$_2$ from a mixture of gases where carbonate is calcined to generate a pure stream of CO$_2$ ready for sequestration. Flue gas from an existing power plant is introduced in the carbonator where the CO$_2$ reacts with CaO to form CaCO$_3$. This process must occur at elevated temperatures (600-650 °C depending on CO$_2$ partial pressure). Removal rates around 80-90% seem to be a reasonable target for this technology.

The formed calcium carbonate is circulated to a different reactor where sorbent regeneration takes place. CaCO$_3$ is calcined and produces a concentrated stream of CO$_2$ suitable for capture and compression. Calcination step is highly energy demanding and it will likely occur at temperatures above 920 °C. Heat requirements for sorbent calcination are covered by oxyfuel combustion in the second reactor itself. Once regenerated, the sorbent is returned to the carbonator to begin a new sorption cycle.

Because of the elevated temperatures, the entire cycle might be integrated in a steam cycle, reducing energy penalties of the capture system by several percentage points. The cost of natural sorbents for these cycles is significantly low, reducing operation costs.

This chapter examines the energy penalties of the Ca-looping CO$_2$ capture system, different types of sorbents and their performance subjected to repeated cycles of carbonation and calcination, the CO$_2$ capture efficiency and the possibility of integration of Ca-looping and power plants to reduce energetic penalties.

Keywords: Calcium looping, carbonation-calcination, high temperature sorbents, lime, cyclic sorbent degradation, energy integration, interconnected CFB, post-combustion, power plant, solid circulation, purge flow, pilot plants.

This chapter introduces the high temperature carbonation/calcination cycle of calcium based compounds to remove CO$_2$ from flue gas, namely Ca-looping

*Corresponding author Luis M. Romeo: CIRCE Institute - University of Zaragoza, C/Mariano Esquillor Gomez, 15, 50018-Zaragoza. Spain; Tel: +34 976 76 25 70; Fax: +34 976 73 20 78; E-mail: luismi@unizar.es

Rosa-Hilda Chavez and Javier de J. Guadarrama (Eds)

process. Carbonation efficiency is limited by the equilibrium of carbonation-calcination reactions. This equilibrium permits capture efficiencies of around 95% for typical flue gases concentrations of CO_2 under process conditions (8-15% vol. depending on the fuel), 650 °C and atmospheric pressure. However, under real operation this maximum value is not always reached since a number of factors influence this efficiency in different ways.

In the light of calcium looping characteristics, the main variables affecting the efficiency of the capture process and its economical feasibility will be detailed. The selection of the sorbent is crucial since mechanical resistance and chemical properties determine the behaviour of the process. Attrition resistant materials are preferred to minimize the loss of the inventory out of the system. Also the cost of the sorbent is a crucial factor to ensure the feasibility of large scale Ca-looping capture plants. The CO_2 carrying capacity of the sorbent and its decay with the increasing number of cycles directly affect the capture efficiency.

A capture plant will always be associated to a power plant or any other big source of CO_2 and one advantage of calcium looping is the possibility of energy integration to limit the energy penalties of the capture system. The thermal efficiency of the global system and the cost of avoided CO_2 strongly depend on the key variables of the Ca-looping process, therefore the optimal ranges for operation parameters must be determined, taking into account the whole integrated system.

1. FUNDAMENTALS ON Ca-LOOPING TECHNOLOGY

This technology makes use of the idea that lime may be reused in a cyclic process to remove CO_2 from a mixture of gases where carbonate is calcined to generate a pure stream of CO_2 ready for sequestration. The carbonation-calcination equilibrium reaction can be expressed as shown in equation 1. Carbonation reaction is an exothermic reaction and the backward reaction is, consequently, endothermic ($\Delta H^{\circ}_{298} = -178$ kJ/mol).

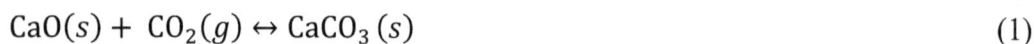

$$CaO(s) + CO_2(g) \leftrightarrow CaCO_3(s) \tag{1}$$

Under specific operating conditions (pressure and temperature), the relationship between the equilibrium CO_2 concentration in the gas after CO_2 capture can be obtained from the equilibrium constant expression [1], presented in equation 2. This expression was calculated using experimental data with temperatures above 896.85 °C and partial pressures of CO_2 above 101 000 Pa [2].

$$\log P_{CO_2} = 7.079 - 8\ 308\ T^{-1} \tag{2}$$

The proposal of using of Ca-based sorbents to remove CO_2 was initially presented by Tessié du Motay and Maréchal [3] back in 1868. The application of the concept was first patented in 1933 and technologically developed later on for sorption-enhanced hydrogen production, the so-called CO_2 Acceptor Gasification Process [4]. In 1976, the first large demonstration pilot plant was built in South Dakota, making use of a high pressure dual fluidized bed gasifier system operating with a coal input up to 10 t/h [5]. The coal was gasified using steam and the endothermic reactions were supported by a hot stream of lime which also served as CO_2 sorbent. None of these preliminary applications of lime for CO_2 removal were focused on carbon capture and storage development.

Silaban and Harrison [6] proposed, in 1995, the use of this cyclic system for CO_2 removal from flue gas with the aim of storage, classified as "hot" post-combustion capture technology. The full CO_2 capture scheme of this application was first proposed by Shimizu *et al.,* [7], who developed a conceptual study on an atmospheric pressure system of two interconnected circulating fluidized beds, also called dual CFB system (dCFB). Circulating fluidized beds provides good gas-solid contact and a uniform temperature, making this technology adequate to build carbonator and calciner reactors.

According to Fig. **1**, the flue gas from an existing power plant (F_{gas}) is introduced in the carbonator where the CO_2 contained in this stream (F_{CO2}) reacts with CaO to form $CaCO_3$. This process must occur at elevated temperatures (600-650 °C depending on CO_2 partial pressure). The extent of the reaction depends on the equilibrium which is given by equation 2, thus CO_2 removal from flue gas stream is limited by operating conditions. Removal rates around 80-90% seem to be a reasonable target for this technology [8]. In second step, the formed calcium carbonate is circulated to a different reactor where sorbent regeneration takes place. $CaCO_3$ is calcined producing a concentrated stream of CO_2 suitable for capture and compression. Calcination step will likely occur at temperatures above 920 °C and in the absence of high concentration of steam. Heat requirements for sorbent calcination are covered by oxyfuel combustion of coal in the second reactor itself. Once regenerated, the sorbent is returned to the carbonator to set out a new sorption cycle. Ca-looping process presents similarities to amine scrubbing post-combustion capture. Both processes require two different stages operating at different conditions for capture and regeneration of the CO_2 sorption material. Regeneration step is highly energy demanding in both cases. The primary difference is that, because of the elevated temperatures of gas streams, the entire cycle might be used as part of a steam cycle reducing energy penalties of the capture system by several percentage points. From the economical point of view, the price of natural sorbents for high temperature cycles is significantly lower than amines, reducing operation costs.

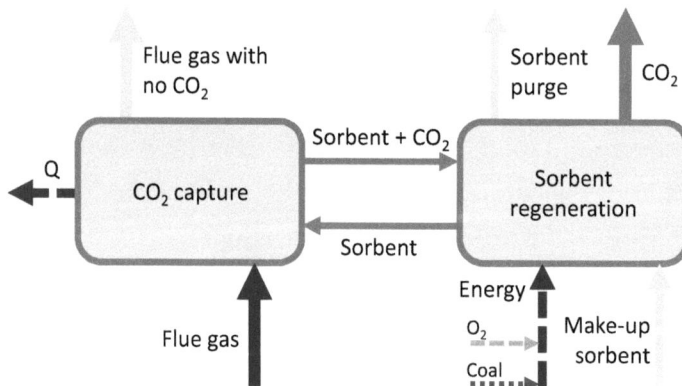

Figure 1: Schematic representation of carbonation-calcination loop.

1.1. Energy Penalties of the Ca-Looping CO$_2$ Capture System

Strong energy penalties are associated not only with the endothermic regeneration of $CaCO_3$ in the calciner reactor but also with the need to raise up the temperature of the solid stream recycled from the carbonator and the fresh limestone added to the calciner, the oxygen separation process for oxyfuel combustion and the compression of captured CO_2.

Regarding the energy consumption in the Ca-looping itself, three main phenomena are to be considered. The calcination stage energy demand will depend on the capture capacity of the plant and the calcination efficiency reached in the calciner. Secondly, a temperature spring of approximately 250 °C must be transferred to the solid stream which circulates between reactors. Finally, the fresh limestone fed into the system must be heated up from ambient temperature to nearly 920 °C and calcined together with the cycled material.

Heat requirements are, therefore, diminished when the circulating solid flow rate decreases because of a lower size of the capture plant or because there is a reduced presence of ashes and sulphur compounds from fuel accumulated in the loop. Large make-up flows improve carbonation conversion, but significant amounts of purge dramatically increase the heat demand at calciner, the oxygen production cost and the auxiliaries consumption.

Higher temperatures in the carbonator or attenuated in the calciner would lower the temperature difference between reactors and, thus, the energy required for solid heating up. These measures which limit energy consumption in the cycle would lead to extremely poor capture efficiencies in the carbonator. There is a confronted influence of every independent operational parameter on the two variables which present the most important effect on the economic feasibility of the Ca-looping technology.

Nevertheless, these parameters do not vary independently since strong interactions exist between them. A deeper knowledge of these interactions and their influence on the effective heat requirements is needed. Rodríguez *et al.,* [9] defined the narrow operating window which must be chosen for a given fuel composition and process characteristics to minimize heat requirements in the calciner. Their model incorporated equilibrium and sorbent performance, reactor data and sulphur/ashes content in the incoming fuel or flue gas. Aiming at a 70% CO_2 capture, they concluded that heat requirements were around 37% when free-ash and free-sulphur fuel were used and no make-up flow was added. Increasing the fresh sorbent addition, heat requirements were found to decrease down to 30% since the improved average activity of the sorbent population, $X_{ave}= 0.23$, diminishes the rate of exchanged solids between reactors. From this value up, further increase of purge makes the heat requirements increase since they become dominated by the calcination of fresh limestone. They concluded that an optimum level of sorbent activity which minimized heat requirements in the calcination stage appears under different operating conditions, processes and fuel characteristics.

Different studies [7, 10, 11] have reported that the resulting energy efficiency of power generation plants when accounting this CO_2 capture process suffers a decrease of around 8-10 percentage points due to Ca-looping inherent needs without considering oxygen production.

Cryogenic distillation is the best suited existing technology for oxygen production rates as those demanded for commercial power plant sizes [12, 13], requiring in most cases two or three parallel ASU [14]. Each unit consists of a number of pressurized distillation columns, in which the rich oxygen stream is produced. After flashing, the generated stream is then purified in a low pressure zone. A train of heat exchangers at low temperature minimizes the thermal energy requirements. The specific power consumption for an ASU approximates values around 220 kWh/t O_2 [12, 14, 15] that nearly amounts a 60% of the total power consumption for CCS in oxyfuel applications and reduces the overall efficiency of the power plant by about 7 - 9 percentage points [14-18]. Latest technology improvements point out an already reached reduction of specific consumption down to 160 kWh/tO_2. Further improvements are expected and target values are fixed in 120 kWh/t O_2.

CO_2 conditioning and compression to 100-120 bar also require a significant amount of power to drive compressors. Energy requirements of these systems are typically 90 - 120 kWh/t CO_2 [19]. For an existing sub-critical power plant with 38% gross efficiency, CO_2 compression demands more than 100 kWh per MWh produced. This value reduces the overall efficiency of the power plant by about 3 - 4 percentage points.

1.2. Sorbent Performance

The application of reaction 1 to CO_2 capture requires that the sorbent is subjected to repeated cycles of carbonation and calcination. One critical aspect involving Ca-based sorbents deals with the decay on conversion efficiency of carbonation reaction under process conditions after several cycles associated with sorbent degradation.

The kinetics of the carbonation reaction suffers a transition from a fast reaction rate to a very slow rate controlled by diffusion through the newly formed product layer [20, 21]. This layer prevents the full carbonation of a particle on a time scale useful for industrial uses. The partial conversion of sorbent at this transition point of the carbonation kinetics is known as the maximum carbonation conversion or carrying capacity, X_N [4, 6, 22]. From a practical point of view, to allow a compact design of the carbonator reactor in Fig. **1**, only the fast reaction period is of interest.

Sintering refers to changes in pore shape, pore shrinkage and grain growth that sorbent particles suffer during extreme heating. It was observed that the longer exposure and higher temperature of calcination step, the more intense sintering of lime, while carbonation reaction had little influence on microstructure changes of lime [23]. Other parameters affecting microscopic structure of the sorbent are high partial pressures of CO_2 and steam and the presence of impurities [24]. The reduction of reactive surface area due to this elimination of microporosity causes a drop-off in reactivity. Disregarding further phenomena such as sulphation or attrition, the decay rate of the maximum carbonation conversion has been extensively modelled in literature [22, 25-27]. Most of these models are semi-empirical and cannot describe the sorbent behaviour from their properties. Although new approaches to predict the conversion decrease, applying the random pore model, were recently developed by Grasa *et al.,* [28].

Grasa and Abanades [26] proposed the most extended semi-empirical equation to model sorbent conversion decay over a large number of cycles, equation 3, which presents very high accuracy. After extended series of experimental tests, they found that there is a residual activity still remaining after 500 cycles. The decay rate constant which also appears in equation 3 takes a different value for every sorbent.

$$X_N = [(1 - X_r)^{-1} + k_{deact}N]^{-1} + X_r \qquad (3)$$

Every natural limestone presents different characteristic long-term carrying capacity and initial conversion depending on the initial pore size distribution, sinterization rate, and impurities in the sample. The residual capacity becomes an important parameter in these cyclic processes when comparing different sorbents for CO$_2$ uptake.

There exist competitive reactions between sulphur compounds from combustion, mainly SO$_2$, and the solid sorbent that deactivates it partially and converts the active sorbent into inert material. The affinity of limestone and lime for sulphur has led to their extended use for flue gas desulphurization systems. Depending on the partial pressure of CO$_2$ in the reactor, direct sulphation, equation 4, or indirect sulphation, equation 5, takes place.

$$CaO(s) + SO_2(g) + 1/2\,O_2(g) \leftrightarrow CaSO_4\,(s) \tag{4}$$

$$CaCO_3(s) + SO_2(g) + 1/2\,O_2(g) \leftrightarrow CaSO_4(s) + CO_2(g) \tag{5}$$

The sulphur content of the fuel burnt in the oxyfuel calciner and the sulphur oxides concentration in the flue gas from the original power plant have a strong effect on the maximum capture capacity of the sorbent that must be accounted. Regeneration of calcium sulphate into lime can be carried out requiring very high temperatures which would damage the microstructure of sorbent for CO$_2$ uptake. Therefore, the newly formed sulphur compound is accumulated into the system and represents irreversible losses of useful sorbent which must be replaced. Fan *et al.,* [29] suggested that reaction with SO$_2$ will be beneficial for the economics of the process since it will avoid the requirement of a separate unit for flue gas desulphurization. Prior to experiment with sulphur side-reactions, also Abanades *et al.,* [30] predicted that the carbonator might achieve effective capture of SO$_2$ avoiding an extra plant for gas cleaning. They justified this probable behaviour on the extremely high molar ratios of Ca-sorbent/SO$_2$ (above 20) which will be present in the real system reactor.

With respect to mechanical stability of the solid particles, Ca-based sorbents again present some problematic issues. González *et al.,* [31] analysed experimental attrition results under carbonation/calcination conditions and concluded that there was a strong decrease in average particle size during the initial cycles of operation. This observation was confirmed by Jia *et al.,* [32] experiments which reported lime attrition taking place over the first couple of calcinations. A critical particle size, below 100 μm, was found to be the final average particle size; no matter what was the initial distribution [31]. Partial sulphation of lime leads to a marked decrease in attrition rates at the expense of sorbent deactivation for CO_2 capture [33].

Although attrition may enhance, in some cases, particle reactivity by continuous removal of the particle surface, consequent elutriation will result in the disappearance of a certain amount of active material which must be continuously fed into the system. Limestone attrition in a circulating fluidized bed may be modelled by using equation 6 [34].

This expression estimates the solid percentage not elutriated from the reactor after every carbonation cycle.

$$a_N = 1 - [k_{att} + (1 + k_{att}) \exp(-N \tau_{att}^{-1})] \dots \tag{6}$$

There is a great number of works conducted to improve sorbent performance upon cycling. This can be done by alteration of the process conditions or by sorbent enhancement. The limestone steam hydration [35, 36], the pretreatment of fresh sorbent by steam [37, 38], the steam reactivation of spent sorbent [35, 39-41], there activation by moist air [42] and the hydration with ethanol/water solutions [43] are examples of the research carried out to upgrade the conversion of calcium carbonate. Steam hydration increases pore volume and surface area [35]. Doped limestones have proved better carbonation results than natural ones [34, 44-47]. Small amounts of Na_2CO_3 [34, 45, 46] and NaCl [45] have been used to increase moderately the carbonation conversion.

Cesium-doped calcium sorbents [44] and CaO/Al_2O_3 [47] have also been tested as high-temperature sorbents and could reach important improvements in the carbonation reaction. Dolomite [48, 49] and synthetic $CaCO_3$ (Precipitated Calcium Carbonate PCC) [50, 51] have been investigated as sorbents, showing a better reactivity than CaO obtained from the calcination of natural calcium carbonate. The reason, again, is the increment of pore volume due to sorbent impurities. Dolomite modified with acetic acid solution has been proposed [43, 52] and it exhibits higher carbonation conversion than modified sorbent due to superior surface area and pore volume. In this case, conversion for cycle number 20 is around 50% (depending on carbonation temperature), instead of 20% for natural limestones. Finally, other sorbents are potassium-based or impregnated potassium-based [53, 54]. The equilibrium of these sorbents is different, and carbonation and regeneration temperatures change being more suitable for low temperature capture cycles.

Most research efforts have been focused on improving the sorbent conversion during the carbonation step assuming that, under high enough temperature conditions, fast and complete calcination will occur during the regeneration. More elevated temperatures would lead to faster degradation of the sorbent and larger energy requirements in the calciner. Therefore, the design temperature in the calciner reactor will always be aimed at the lowest value which ensures effective regeneration of limestone with minimum energy requirements. This target temperature is around 900 °C and, at these low temperatures, the kinetics of the calcination reaction may not be sufficient to achieve the required elevated calcination efficiency. Martínez *et al.,* [55] have recently devoted a thoughtful study to develop a kinetic model that predicts calcination rate under realistic operating conditions. On the basis of their results, it is shown that calcination temperatures between 820-920 °C could be adequate to achieve nearly complete regeneration at typical solid residence time of CFB calciner reactors (2-3 min).

1.3. CO$_2$ Capture Efficiency

The maximum sorbent conversion is declined by the cyclic decay of the sorbent and other negative factors, such as attrition or sulphation competing side-reactions. Make-up flow of fresh limestone, F_0, will be required to compensate for the sorbent degradation, purged flow to avoid inert accumulation, and elutriation of fines. Fresh sorbent flow keeps an adequate level of average carrying capacity of solid population in the reactor. Hence, sorbent durability or reactivation techniques and mechanical resistance may reduce make up costs and will deserve an independent analysis.

As a consequence of the requirement for a constant introduction of fresh material in the system and a continuous removal of spent sorbent, the solid population within the loop will present a distribution of cyclic ages. The mass fraction of particles, r_N, entering the carbonator in the recycled regenerated sorbent stream, F_R, that have circulated N times through the loop is calculated from a succession of mass balances. The composition of the solid stream returning to the carbonator reactor is mainly composed of CaO, CaSO$_4$ and ashes. The final expression of r_N will depend on the chosen configuration, *i.e.* number of reactors, interconnections between them, possible solid recirculation into the same fluidized bed or purge location. For the proposed dCFB system presented in figure1, the mass fraction of particles that has experienced N calcinations is given by equation 7.

$$r_N^* = F_0 F_R^{-1}(1 + F_0 F_R^{-1})^{-N} \tag{7}$$

To preserve the stationary mass balance in the system, the molar flow of Ca-compounds in the stream of purged ashes and spent sorbent must be equal to the fresh limestone molar flow (F_0). Thus, the ratio F_0/F_R may be used to define the magnitude of the purged flow as a percentage. The solid purge in the system, f_p, has been defined as a percentage over the total mass flow leaving the reactor when no purge is considered. This output flow takes into account all inlet flows and the corresponding chemical reactions within each reactor.

Since the conversion of each of these populations of particles is only determined by the number of cycles, the maximum average conversion of the total population in the carbonator is given by equation 8.

$$X_{ave} = \sum_{N=1}^{\infty} r_N^* X_N \tag{8}$$

The expression of X_{ave} for the circulating lime is a function of the limestone carbonation capacity, the percentage of purged solid, f_p, the stoichiometry of the sorption reaction and the location of the purge. Once the stoichiometry, the make-up flow, and the internal solid circulation in the system are set, the maximum average conversion of the solid population can be calculated.

Maximum average capture capacity of the sorbent (X_{ave}) as expressed in equation 8 only takes into account the degradation experienced by sintering. However, real average sorption capacity is expected to be lower than the maximum average capture capacity. Actually, ashes derived from coal combustion in the calciner interact with sorbent particles, reducing the amount of available active sorbent in the loop and reactor design (residence time and contact regime) influence kinetics. Experimental results presented by Abanades *et al.,* [56] show that real conversion varies from 70 to 80% of the maximum theoretical value. When sorbent deactivation by $CaSO_4$ formation is accounted, all the sulphur contained in the fuel leaves the system in the solid purge. Considering the conservative assumption that sulphur only reacts with the "active" part of CaO for CO_2 capture, Rodríguez *et al.,* gave a more restrictive value for the average capture capacity [57], equation 9.

$$X_{ave} = \sum_{N=1}^{\infty} r_N^* X_N - F_{CO_2}\left(F_0 r_{C/S}\right)^{-1} \tag{9}$$

From a simplified closure of the carbon mass balance in the carbonator, the relationships between the maximum achievable CO_2 captured flow and the Ca-based sorbent flow shown in equation 10 may be defined.

$$F_{CO_2 captured} = F_{CO_2} \eta_{carb} = F_{Ca} X_{ave} \tag{10}$$

The maximum achievable CO_2 capture efficiency is therefore calculated in a simplified form as shown in equation 11 where R represents the molar ratio between calcium oxide introduced in the carbonator and CO_2 in the flue gas stream, the so-called Ca-looping ratio.

$$\eta_{carb} = R X_{ave} \tag{11}$$

Low sorbent conversion may be compensated by increasing the Ca-looping ratio to achieve a certain level of CO_2 capture efficiency from the gas phase. Large values of R directly increase the carbonation reaction efficiency but also energy penalties due to larger sorbent circulation and calcination heat requirements. A recent modelling work of Charitos *et al.,* [58] has shown the actual capture efficiency to lie between 30% and above 90% based on carbonator active space time variation. They applied the fundamental carbonator reactor design equation to close the carbon mass balance instead of using equation 10. Equation12 relates capture efficiency to the calcium inventory in the reactor and the average reaction rate of these compounds.

$$F_{CO_2}\eta_{carb} = n_{Ca}(dX_{carb}/dt)_{reactor} \tag{12}$$

Two different approaches are derived to describe the efficiency of carbonation reaction by means of all the parameters in the system.

$$\eta_{carb} = k_S\varphi\tau(X_{ave} - X_{carb})\left(\overline{v_{CO_2}} - v_{eq}\right) \tag{13}$$

$$\eta_{carb} = k_S\varphi f_{active}\tau X_{ave}\left(\overline{v_{CO_2}} - v_{eq}\right) \tag{14}$$

The first approach assumes that all the particles in the bed are able to react with CO_2 with the same average characteristics, leading to the expression given in equation 13. The second approach was derived by Alonso *et al.,* [59] and considers that only a fraction of lime particles in the carbonator are active to react under the fast kinetics regime, equation 14. Results obtained from both expressions have been validated with experimental data from *INCAR-CSIC* and *Stuttgart University* facilities showing very good agreement.

2. ENERGY INTEGRATION OF Ca-LOOPING AND POWER PLANTS

Energy consumptions in the capture cycle penalize by several percentage points the energy efficiency of the power plant. One of the key advantages of Ca-looping is the potential for retrofitting existing power plants or other stationary industrial CO_2 sources [8, 9]. Because this process is being operated at relatively high temperatures (around 600-650 °C), the majority of the energy input can be recuperated from the hot gas and solid streams exiting the system. Also, the heat generated by the exothermic carbonation reaction which must be evacuated to control operating temperature is available for integration. Low-grade heat may be recuperated from the CO_2 intercoolers included in the compression stage and the ASU.

The cryogenic plants used for O_2 separation have been successfully integrated with IGCC technology [60], but the potentiality for a proper energetic integration of an ASU in oxyfuel systems presents some restrictions. The possibilities to improve the efficiency of the system by using the low temperature heat from the ASU intercoolers are limited. Different options to integrate the available energy within the power plant are found in the literature; Darde *et al.,* [61] proposed to heat the water after condensation in the steam cycle and other research groups [62, 63] to increase O$_2$ temperature before the boiler.

Again, the potential reduction of penalties associated with compression stage is subjected to the use of low temperature heat from the CO$_2$ intercoolers. This energy is commonly integrated for feed-water preheating, reducing the steam turbine bleeds and increasing the net power output and power plant efficiency.

For oxyfuel calciner in Ca-looping CCS systems, energy penalties associated with ASU and CO$_2$ compression would reduce power plant efficiency in 11-13 percentage points, representing more than 95% of parasitic losses. By integration of the waste heat in the steam cycle to preheat feed-water, the efficiency penalty in cold geographical regions [12, 62] where low ambient temperature allows to maximize the use of this heat, may drop-off down to 10-12 points. Nevertheless, it is not possible to make use of all the available energy and the residual heat must be dissipated in an additional cooling system. A completely new integration approach is introduced by Romeo *et al.,* [64] studying the possibility of generating extra power through an organic Rankine cycle run with the waste heat from O$_2$ separation and CO$_2$ compression stages.

The largest saving potential is linked to high temperature flows integration with the power plant. High-temperature heat flows from sorption-desorption processes are carbonation reaction heat, energetic content of gases leaving both reactors and solids purge stream. According to the heat exchangers temperature levels, the energetic value of these streams is employed to design the high-pressure equipments of a supercritical steam cycle; *i.e.* heat recovery steam generator, reheater and high-pressure pre-heaters. Romeo *et al.,* [63] computed an efficiency penalty of only 4-5 percentage points when heat recuperation into a supercritical steam cycle was considered. Nevertheless, this study does not assess the influence of main operational variables of the capture cycle in the efficiency of the power plant.

Rodríguez *et al.,* [9] modelled the Ca-looping to define the conditions which minimize heat requirements in the calciner. When waste heat from the capture cycle is integrated into the power plant, it is important to notice that the best possible scenario to minimize energy penalties in power plant efficiency may not be coincident with the scenario which minimizes heat requirements in the calciner.

Further studies have been developed to analyse the energetic performance and economics of the integrated system of a Ca-looping process and an existing coal-fired electric generation power plant [65]. As pointed out, energy penalties in Ca-looping are affected by many different variables which, at the same time, have an influence on them. An optimum operational point from the energetic point of view of the single capture plant may not be coincident with the optimal conditions to minimize the cost of captured tonne of CO_2 once the Ca-looping system is integrated with a power plant.

Romeo *et al.,* [65] presented the first sensitivity analysis to assess the influence of solids purge flow, solid circulation ratio and the location of the purge on the total costs of electricity and the cost of captured tonne of CO_2. The target was to find the operating window which yields to the optimal scenario. Mass and energy balances in the system are solved and the obtained results used as input data in the economical model which will evaluate the effect of relevant operational parameters of the integrated system on avoided CO_2 cost.

The cost of avoided tonne of CO_2 is defined in equation 15. This expression is given by the IPCC and takes into account the cost of reducing atmospheric CO_2 emissions while providing the same amount of product as the reference plant without CO_2 capture system [66].

$$Cost\ of\ avoided\ t\ CO_2 \\ = [(COE)_{capture} - (COE)_{ref}][(CO_2 \times kWh^{-1})_{ref} - (CO_2 \times kWh^{-1})_{capture}]^{-1} \quad (15)$$

The basic concept of the integrated system is to apply the Ca-looping CO_2 capture technology to clean the flue gas from an existing power plant and to integrate the available energy flows from capture and compression processes into a new supercritical steam cycle which generates extra electrical power. Fig. **2** schematically represents the three main blocks included in the model; reference plant, capture cycle/compression train and supercritical steam cycle.

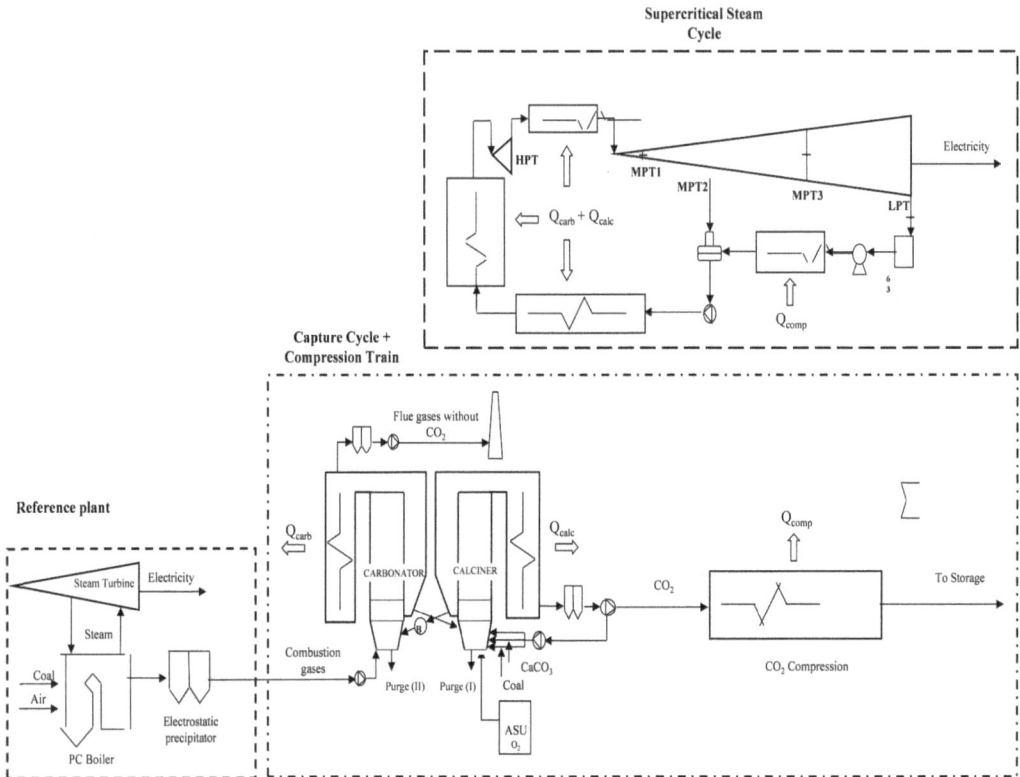

Figure 2: Integration of residual heat streams from Ca-looping into a new steam cycle.

The reference power plant, used as baseline case, generates 500 MW$_e$ and the net efficiency of the unit is 38.11% LHV. It is considered for calculations a high-rank coal whose sulphur content remains below 0.65% dry basis. Flue gas from the reference power plant is fed to the Ca-looping with a CO$_2$ content of 21.72 wt%. Recirculation of the CO$_2$ concentrated gas from calciner is essential in oxyfuel combustor performance to provide enough gas velocity to reach CFB operation and to control bed temperature. This flow ranges from 17 to 21% of the total outlet stream ensuring an O$_2$ mass fraction in the inlet gas flow of 30%. A typical value of 220 kWh/t O$_2$ is considered to estimate the ASU power consumption [66].

The supercritical steam cycle (see Table **1** for steam properties) consists of a heat recovery steam generator, high pressure turbine, reheater, three medium-pressure turbines and one low-pressure turbine, condenser, low-pressure preheating system, deaerator, high-pressure feed water heaters and economiser.

Main technical assumptions considered for this simulation are gathered in Table **1**. Steam mass flow production mainly depends on solid circulation in the Ca-looping so the design and extra power output vary depending on the same variables which affect capture efficiency; *i.e.* Ca-looping ratio and purge percentage.

The energy integration with CO_2 compression train makes unnecessary the steam turbine bleeds for low-pressure water preheating prior the deaerator, although high-pressure pump is driven by a MPT_2 turbine extraction.

Table 1: Main technical assumptions of the model

Ca-looping Capture System	
Purge final temperature	180 °C
Carbonation final flue gas temperature	180 °C
Calciner flue gas temperature before compression	53-59 °C
CO₂ Compression Train	
Intercooling temperature at CO_2 compression	50 °C
Pressure ratio at CO_2 compression	3.3
CO_2 recirculation	17%-21%
ASU consumption	220kWh/tO_2
CO_2 properties after compression	80 °C, 120 bar
Supercritical Steam Cycle	
Power plant boiler flue gas temperature	180 °C
Deaerator pressure	7 bar
Condenser pressure	0.045 bar
Live steam properties at turbine inlet	600 °C, 290 bar
Reheat steam properties at turbine inlet	620 °C, 48.5 bar

The total compression energy required for CO_2 conditioning, 120 bar and 80 °C, represents about 5-14% of the total energy output. The total energy output accounts for the reference plant and the new steam cycle power generation. The compression process requires intercooling stages to reduce energetic demands and to avoid an excessive increase in CO$_2$ temperature. A potentially recoverable low-temperature heat, Q_{comp}, must be evacuated.

The extra heat released by the carbonation exothermic reaction and the sensible heat recuperated from clean gas leaving the carbonator at high temperature, Q_{carb}, will be allocated in the new steam cycle. Recovered heats from carbonator, Q_{carb}, and calciner, Q_{calc}, are used to design the high-pressure equipments of a supercritical cycle according to the heat exchangers temperature levels. Heat recovery steam generators take advantage of the flue gas from carbonator at 650 °C and from calciner at 930 °C. The sensible heat of the solid purge stream whose composition is different whether it is performed in one or the other reactor, is used to reheat both CO$_2$ recirculation to the calciner and O$_2$ entering the calciner. A solid-gas heat exchanger is used to reduce the temperature of this stream from the corresponding temperature operation of the purged reactor down to 180 °C. Intercooling CO$_2$ compression heat, Q_{comp}, is used in low-pressure heat exchangers in the condensate section of the steam cycle.

Obtained cost of avoided CO$_2$ seems to be competitive with the costs provided by other more mature capture technologies. Values below 15 €/tCO$_2$ are found under simulated scenarios. Although, this study represent a simplified and preliminary evaluation and further information must be accounted in the thermoeconomical model to obtain more reliable figures; it is useful to point out the potentiality of this technology after energy integration against other capture processes.

It may be concluded that the target values for operating parameters once the Ca-looping process is integrated with a new power plant are found at Ca-looping ratios around 4-5 and purge percentages around 2%. These values may be translated into actual net solid circulation flows and mass purge flows for the design of large-scale real plants. For a reference plant of 500 MWe fed by high quality coals, net solid circulations around 10 kg/m^{2}s and purge flows of 25 kg/s seem to be in the nearby of the optimum scenario. The influence of coal sulphur content on operating parameters is of importance. Increasing ten times the sulphur content of the coal, the required net solid circulation to optimize the performance of the system is duplicated for low purges and increased in approximately a 25% for large purge values.

3. Ca-LOOPING EXPERIMENTAL PLANTS

Although mathematical modelling and simulation are precious tools to understand the theoretical behaviour of any process, experimental experience is always required in the development chain of a new technology. Strong efforts are being done worldwide to demonstrate the technical feasibility of Ca-looping capture process at pilot and large scale. The existing facilities developed under different projects are described and the results obtained briefly introduced in the following. These projects, up to 120 kW_{th}, are located all around the world.

The INCAR (*Instituto Nacional del Carbón*) is a division of the Spanish Research Council (CSIC) located in Oviedo (Spain). INCAR-CSIC developed the first test facility comprising two interconnected CFB reactors. The current thermal power of this pilot plant is 30 kW_{th}. Continuous experiments for several hours have been performed using air-fired calcination at 800-900 °C. CO_2 capture efficiencies were consistent and stable between 70% and 90%, depending on the make-up flow of sorbent which leads to maximum average activity of limestone population, X_{ave}, around 0.2-0.3. Most of the attrition was found to occur over the first calcination stage and therefore fresh limestone was added to keep the mass balance. Once the material had been calcined, solids presented a more or less uniform particle size below 100 µm and attrition appeared to stabilise on continued operation [56, 67]. The same test plant was used to validate the *in-situ* CO_2 capture using CaO from biomass combustion in the carbonator itself [56]. A CO_2 capture efficiency higher than 70% was obtained with sufficiently high solids circulation rates of CaO and solids inventories within the carbonator which operated at 700 °C.

The IFK (*Institut für Feuerungs- und Kraftwerkstechnik*), in the *University of Stuttgart*, has a 10 kW_{th} pilot plant consisting of a CFB carbonator and BFB calciner operated with air. The control of solid looping rate between the beds was performed by a cone valve [68, 69]. This group has extensively studied the hydrodynamic of the apparatus. A parametric study reported CO_2 capture efficiency of over 90%. The CO_2 capture efficiency was found to decrease with decreasing Ca-looping ratio and space time (number of moles of CaO in carbonator/CO_2 flow to carbonator) [70].

The *Vienna University of Technology* has investigated the 'Adsorption Enhanced Reforming' process development with the main target of enhancing the steam gasification of biomass. The system consists of a dual fluidized bed system with a gasifier/carbonator operating at 600-700 °C at atmospheric pressure and a combustor/calciner, 100 kW$_{th}$ fuel power. Much work has been done on mechanical properties and reactivity of sorbents for the process [71, 72]. AER operation increased H$_2$ production from biomass from 40% to 75% (vol.) compared to an analogous already commercial process without CO$_2$ sorbent. The process has been subjected to larger experimental tests in an 8 MW$_{th}$ fuel input CHP unit in Güssing, Austria [73].

The *Canmet Energy and Technology Centre* is a division of Natural Resources Canada located in Ottawa (Canada) where a 75 kW$_{th}$ dual fluidized bed system setup consisting of two reactors (CFB calciner and BFB carbonator) was designed and operated [74]. Lu *et al.,* reported continuous operating experience with three different calciner operating modes: electrically heated, oxy-combustion with biomass and oxy-combustion with bituminous coal [75]. CO$_2$ capture of around 95% was achieved in the first cycles of operation, dropping to 71% after 25 cycles. This decay in sorbent reactivity is influenced by the strong attrition found in the system.

A variation of the Ca-looping is the carbonation/calcination reaction process developed at *The Ohio State University*. In this process, the CaO/Ca(OH)$_2$ is injected into an entrained bed reactor where it reacts with CO$_2$ and SO$_2$ between 450 and 650 °C; it is then calcined at a high temperature between 850 and 1300 °C. There is a third unit, the hydrator, used to reactivate the sorbent. This cyclic process has been successfully tested at their 120 kW$_{th}$ pilot plant utilizing coal and natural gas mixture as the feed stock. The pilot facility consists of a stoker furnace, whose flue gas is fed into an entrained bed carbonator operating at 500-625 °C, the solids which are passed into an electrically heated rotary calciner operating at 980 °C. The calcined sorbent was removed, hydrated off line, and passed to the flue gas duct [76]. CO$_2$ capture efficiencies above 90% and SO$_2$ capture efficiencies of approximately 100% were achieved.

The department of Thermal Engineering in *Tsinghua University*, China, designed and built a 10 kW$_{th}$ dual fluidized bed system which consists of two BFB reactors constructed to demonstrate the process feasibility of continuous CO$_2$ capture from flue gas [77]. The sorbent particles successfully circulated between the carbonator and the regenerator at high temperatures and the CO$_2$ in the flue gas was continuously removed by the Ca-based sorbent, dolomite, with a high CO$_2$ capture efficiency, around 95%. Optimum range of operating temperatures in the carbonator was 600-680 °C. About 70.4% of CaO was converted to CaCO$_3$ in the carbonator. In the regenerator, the CaCO$_3$ was not completely calcined and more than 13.9wt% CaCO$_3$ was still present in the sorbent flow leaving the regenerator. Fang *et al.,* [77] also highlighted the strong influence of sorbent fragmentation and attrition.

In *Cranfield University*, a twin fluidized bed carbonator/calciner system is used to study different aspects of the Ca-looping process. A 4.5 m high and 10 cm wide CFB carbonator and a 16.5 cm wide BFB calciner were built. Solids are fed from the top of the vessels by using screw feeders. The carbonator is fluidized with real flue gas coming from a 25 kW natural gas burner and synthetic gases and the calciner is oxy-fired with natural gas.

Scaled-up Pilot Plants

In addition to existing pilot plants, there are several large projects (up to 2 MW$_{th}$).

i)　The *University of Stuttgart* has already designed and built a larger 200 kW$_{th}$ facility comprising a dual CFB Ca-looping system [78, 79]. The target of the system is to demonstrate capture efficiency above 90% while self-maintaining temperatures through combustion reactions in the same process rather than using electrical furnaces as used in smaller plants. This plant has a flexible design to operate under realistic conditions with a wide range of solid circulation and fresh flow rates. Long-term runs will provide useful information for further scale-up and economic evaluation of this technology. The findings from the successful demonstration of the calcium looping process will be used for design of a 20 MW$_{th}$ demonstration plant.

ii) Within the *CENIT CO$_2$ Project* a Spanish consortium of companies and research centres has designed and built the first 300 kW$_{th}$ pilot plant using Spanish technology. The pilot plant which burn biomass *in-situ* using the Ca-looping process is located at *Gas Natural La Robla* thermal power station (León, Spain).

iii) *The Institut Energie systeme und Energietechnik* of the *Technical University of Darmstadt* (Germany) with the collaboration of ALSTOM has constructed a 1 MW$_{th}$ test unit design which is expected to be flexible enough to demonstrate fully commercial stage of the Ca-looping and chemical looping combustion. The system consists of two 10 m-height circulating fluidized beds.

iv) A 1.7 MW$_{th}$ pilot plant has been built in the Hunosa 50 MW$_e$ coal power plant CFB of *La Pereda*, using a side stream of flue gas of the commercial plant. The facility consists of two interconnected 15m-height circulating fluidized beds and is designed to treat up to 2600 m^3/h flue gas flow with a capture capacity of 8 tCO$_2$ per day and capture efficiencies around 90%. Commissioning tests started in 2012 and the test campaign was programmed to last one more year. The project focuses on the experimental pilot testing and scaling up of the process at scales in the 1 MW$_{th}$ range.

If these projects are successful in demonstrating the technology, large pilot plant demonstrations on the 20-25 MW$_e$ scale can be expected and full commercial demonstrations in the next 10 to 20 years.

CONCLUSION

Ca-looping capture cycle presents several potential benefits as a CO$_2$ capture process post combustion application mainly associated to the cheap sorbent and the low energy penalty. The energy penalty to the overall efficiency of the associated power plant comes from several sources; 8-10% for the heat requirements of the Ca-looping itself, 7-9% for the oxygen production and around 3-4% for the compression and conditioning of the CO$_2$. The main drawback of the process is the cyclic degradation of the sorbent capacity although it is controllable by modifying operational parameters. Further research work into both keeping the reactivity of the sorbent and properly regenerating the sorbent is being carried out. The results will directly lead to an increase of the capture efficiency of the process.

Ca-looping process requires careful energy integration with the steam cycle of a power station if it is to be incorporated as a post combustion capture process inexpensively. Regarding the results obtained after assessing the energy integration of the Ca-looping with a power plant, the total energy penalty is reduced to around 10-12% and the achieved cost of avoided CO_2 seems to be competitive with more mature capture technologies such as amine scrubbing. Values below 15 €/t CO_2 were found in preliminary studies under different scenarios.

It may be concluded that the target values for operating parameters once the Ca-looping process is integrated with a new power plant are found at Ca-looping ratios around 4-5 and purge percentages around 2%. These values may be translated into actual net solid circulation flows and mass purge flows for the design of large-scale real plants. For a reference plant of 500 MW fed by high quality coals, net solid circulations around 10 kg/m^2s and purge flows of 25 kg/s seem to be in the nearby of the optimum scenario. The influence of coal sulphur content on operating parameters is also of importance. Increasing ten times its content in coal, the required net solid circulation which would optimize the performance of the global system was duplicated for low purges and increased in approximately a 25% for large purge values.

Multiple demonstration projects are currently being conducted and test the process at the scale of 1–2 MW_{th}. In the next years, projects on the order of 10 MW_{th} are planned with an expected fast development to commercial demonstrations. In the future, the role of Ca-looping for other purposes such as enhanced H_2 production may become more important, with savings in terms of overall energy consumption and generated CO_2. The versatility of the cycle together with the low cost of the raw materials used, point to widespread application in the future.

ACKNOWLEDGEMENTS

Declared none.

CONFLICT OF INTEREST

The authors confirm that this chapter contents have no conflict of interest.

REFERENCES

[1] Baker, E.H. The calcium oxide-carbon dioxide system in the pressure range 1-300 atmospheres. *J. Chem. Soc.*, **1962**, *70*, 464-470.

[2] Blamey, J.; Anthony, E.J.; Wang, J.; Fennell, P.S. The calcium looping cycle for large-scale CO₂ capture. *Prog. Energ. Combust.*, **2010**, *36*(2), 260-279.

[3] Tessié du Motay, C.R.; Maréchal, C.R. Industrial preparation of hydrogen. *Bulletin Mensuel de la Societé Chimique de Paris*, **1868**, *9*, 334-334.

[4] Curran, G.P.; Fink, C.E.; Gorin, E. CO₂ acceptor gasification process. Studies of acceptor properties. *Adv. Chem. Ser.*, **1967**, *69*, 141 – 161.

[5] McCoy, D.C.; Curran, G.; Sudbury, J.D. CO₂ acceptor process pilot plant. In: *Proceedings of the 8th Synthetic Pipeline Gas Symposium*, **1976**.

[6] Silaban, A.; Harrison, D.P. High temperature capture of carbon dioxide characteristics of the reversible CaO (*s*) and CO₂ (*g*).*Chem. Eng.Commun.*,**1995**, *137*, 177-190.

[7] Shimizu, T.; Hirama, T.; Hosoda, H.; Kitano, K.; Inahaki, M.; Tejima,K. A twin fluid-bed reactor for removal of CO₂ from combustion processes. *Chem. Eng. Res. Des.*, **1999**, *77*, 62-68.

[8] Dean, C.C.; Blamey, J.; Florin, N.H.; Al-Jeboori, M.J.; Fennell, P.S. The calcium looping cycle for CO₂ capture from power generation, cement manufacture and hydrogen production. *Chem. Eng. Res. Des.*, **2011**, *89*, 836-855.

[9] Rodríguez, N.; Alonso, M.;Grasa, G.;Abanades, J.C. Heat requirements in a calciner of CaCO₃ integrated in a CO₂ capture system using CaO.*Chem. Eng. J.*, **2007**, *138*, 148-154.

[10] Abanades, J.C.; Rubin, E. S.; Anthony, E.J. Sorbent cost and performance in CO₂ capture systems, *Ind. Eng. Chem. Res.*, **2004**, *43*, 3462-3466.

[11] Abanades, J.C.; Anthony, E.J.; Wang, J.; Oakey, J.E. Fluidized bed combustion systems integrating CO₂ capture with CaO. *Environ. Sci. Technol.*, **2005**, *39*, 2861-2866.

[12] Andersson, K.; Johnsson, F. Process evaluation of an 865 MWe lignite red O₂/CO₂ power plant. *Energy Convers. Manage.*, **2006**, *47*, 3487-3498.

[13] Andersson, K.; Johnsson, F.; Strömberg, L. Large scale CO₂ capture: Applying the concept of O₂/CO₂ combustion to commercial process data. *VGB powertech*, **2003**, *83*, 1-5.

[14] Toftegaard, M. B.; Brix, J.; Jensen, P. A.;Glarborg, P.; Jensen, A. D. Oxy-fuel combustion of solid fuels. *Prog. Energ.Combust.*, **2010**, *36*(5), 581-625.

[15] Amann, J. M.;Kanniche, M.;Bouallou, C. Natural gas combined cycle power plant modified into an O₂/CO₂ cycle for CO₂ capture. *Energy Convers. Manage.*, **2009**, *50(3)*, 510-521.

[16]   Beér, J.M. High efficiency electric power generation: The environmental role. *Prog. Energ. Combust.*, **2007**, *33(2)*, 107-134.

[17]   Davison, J. Performance and costs of power plants with capture and storage of CO_2. *Energy*, **2007**, *32(7)*, 1163-1176.

[18]   Singh, D.; Croiset, E.; Douglas, P.L.; Douglas, M.A. Techno-economic study of CO_2 capture from an existing coal-fired power plant: MEA scrubbing *vs.* O_2/CO_2 recycle combustion. *Energy Convers. Manage.*, **2003**, *44(19)*, 3073-3091.

[19]   Aspelund, A.;Jordal, K. Gas conditioning: The interface between CO_2 capture and transport. *Int. J. Greenh. Gas Con.*, **2007**, *1(3)*, 343-354.

[20]   Alvarez, D.; Abanades, J.C. Determination of the critical product layer thickness in the reaction of CaO with CO_2. *Ind. Eng. Chem. Res.*, **2005**, *44(15)*, 5608-5615.

[21]   Bhatia, S.K.; Perlmutter, D.D. Effect of the product layer on the kinetics of the CO_2-lime reaction. *AIChE J.*, **1983**, *29(1)*, 79-86.

[22]   Abanades, J.C. The maximum capture efficiency of CO_2 using a carbonation/calcination cycle of $CaO/CaCO_3$.*Chem. Eng. J.*, **2002**, *90*, 303-306.

[23]   Sun, P.; Grace, J.R.; Lim, C.J.; Anthony, E.J. The effect of CaO sintering on CO_2 capture in energy systems. *AIChE J.*, **2007**, *53*, 2432-2442.

[24]   Borgwardt, R.H. Sintering of nascent calcium oxide. *Chem. Eng. Sci.*, **1989**, *44*, 53-60.

[25]   Abanades, J.C.;Alvárez, D. Conversion limits in the reaction of CO_2 with lime. *Energy Fuels*, **2003**, *17*, 308-315.

[26]   Grasa, G.S.; Abanades, J.C. CO_2 capture capacity of CaO in long series of carbonation/calcination cycles. *Ind. Eng. Chem. Res.*, **2006**, *45(26)*, 8846-8851.

[27]   Wang, J.S.; Anthony, E.J. On the decay behaviour of the CO_2 absorption capacity of CaO-based sorbents. *Ind. Eng. Chem. Res.*, **2005**, *44*, 627-629.

[28]   Grasa, G.; Murillo, R.; Alonso, M.; Abanades, J.C. Application of the random pore model to carbonation cyclic reaction. *AIChE J.*, **2009**, *55*, 1246-1255.

[29]   Fan, L.S.; Lin, F.X.; Ramkumar, S. Utilization of chemical looping strategy in coal gasification processes. *Particuology*, **2008**, 6, 131-142.

[30]   Abanades, J.C.; Anthony, E.J.; Lu, D.Y.; Salvador, C.; Alvarez, D. Capture of CO_2 from combustion gases in a fluidized bed of CaO. *AIChE J.*, **2004**, *50*, 1614-1622.

[31]   González, B.; Alonso, M.; Abanades, J.C. Sorbent attrition in a carbonation/calcination pilot plant for capturing CO_2 from flue gases. *Fuel*, **2010**, *89(10)*, 2918-2924.

[32]   Jia, L.; Hughes, R.; Lu, D.; Anthony, E.J.; Lau, I. Attrition of calcinin glimestones in circulating fluidized-bed systems. *Ind. Eng. Chem. Res.*, **2007**, *46(15)*, 5199-5209.

[33]   Lu, D.Y.; Hughes, R.W.; Anthony, E.J.; Manovic, V. Sintering and reactivity of CaCO3-based sorbents for *in situ* CO_2 capture in fluidized beds under realistic calcination conditions. *J. Environ. Eng.-ASCE*, **2009**, *135 (6)*, 404-410.

[34]   Fennell, P.S.; Pacciani, R.; Dennis, J.S.; Davidson, J.F.; Hayhurst, A.N. The effects of repeated cycles of calcination and carbonation on a variety of different limestones as measured in a hot fluidized bed of sand. *Energy Fuels*, **2007**, *21*, 2072-2081.

[35]   Manovic, V.; Anthony, E.J. Sequential SO_2/CO_2 capture enhanced by steam reactivation of a CaO-based sorbent. *Fuel*, **2008**, *87*, 1564-1573.

[36]   Manovic, V.; Lu, D.; Anthony, E.J. Steam hydration of sorbents from a dual fluidized bed CO_2 looping cycle reactor. *Fuel*, **2008**, *87*, 3344-3352.

[37]   Hughes, R.W.; Lu, D.; Anthony, E.J.; Wu, Y. Improved long-term conversion of limestone-derived sorbents for *in situ* capture of CO_2 in a fluidized bed combustor. *Ind. Eng. Chem. Res.*, **2004**, *43*, 5529-5539.

[38] Zeman, F. Effect of steam hydration on performance of lime sorbent for CO$_2$ capture. *Int. J.Greenh. Gas Con.*, **2008**, *2(2)*, 203-209.

[39] Manovic, V.; Anthony, E.J. SO$_2$ retention by reactivated CaO-based sorbent from multiple CO$_2$ capture cycles. *Environ. Sci. Technol.*, **2007**, *41*, 4435-4440.

[40] Manovic, V.; Anthony, E.J. Steam reactivation of spent CaO-based sorbent for multiple CO$_2$ capture cycles. *Environ. Sci. Technol.*, **2007**, *41*, 1420-1425.

[41] Sun, P.; Grace, J.R.; Lim, C.J.; Anthony, E.J. Investigation of attempts to improve cyclic CO$_2$ capture by sorbent hydration and modification. *Ind. Eng. Chem. Res.*, **2008**, *47*, 2024-2032.

[42] Fennell, P.S.; Davidson, J.F.; Dennis, J.S.; Hayhurst, A.N. Regeneration of sintered limestone sorbents for the sequestration of CO$_2$ from combustion and other systems. *J. Energy. Inst.*, **2007**, *80*, 116-119.

[43] Li, Y.J.; Zhao, C.S.; Qu, C.R.; Duan, L.B.; Li, Q.Z.; Liang, C. CO$_2$ capture using CaO modified with ethanol/water solution during cyclic calcination/carbonation. *Chem. Eng. Technol.*, **2008**, *31*, 237-244.

[44] Roesch, A.; Reddy, E.P.; Smirniotis, P.G. Parametric study of Cs/CaO sorbents with respect to simulated flue gas at high temperature. *Ind. Eng. Chem. Res.*, **2005**, *44*, 6458-6490.

[45] Salvador, C.; Lu, D.; Anthony, E.J.; Abanades, J.C. Enhancement of CaO for CO$_2$ capture in a FBC environment. *Chem. Eng. J.*, **2003**, *96*, 187-195.

[46] Seo, Y.; Jo, S.H.;Ryu, C.K.; Yi, C.K. Effects of water vapour pretreatment time and reaction temperature on CO$_2$ capture characteristics of a sodium-based solid sorbent in a bubbling fluidized-bed reactor. *Chemosphere*, **2007**, *69*, 712-718.

[47] Wu, S.F.; Li, Q.H.; Kim, J.N.; Yi, K.B. Properties of a nano CaO/Al$_2$O$_3$CO$_2$ sorbent. *Ind. Eng. Chem. Res.*, **2008**, *47(1)*, 180-184.

[48] Chrissafis, K.; Paraskevopoulos, K.M. The effect of sintering on the maximum capture effiency of CO$_2$ using carbonation/calcination cycle of carbonate rocks. *J. Therm. Anal. Calorim.*, **2005**, *81*, 463-468.

[49] Dobner, S.; Sterns, L.; Graff, R.A.; Squires, A.M. Cyclic calcination and recarbonation of calcineddolomite. *Ind. Eng. Chem. Process. Des. Dev.*, **1977**, *16,* 479-486.

[50] Gupta, H.;Iyer, M.V.;Sakadjian, B.B. Reactive separation of CO$_2$ using pressure palletized limestone. *Journal of Environmental Technology Management*, **2004**, *4*, 3-20.

[51] Sakadjian, B.B.;Iyer, M.V.; Gupta, H.; Fan, L.S. Kinetics and structural characterization of calcium-based sorbents calcined under subatmospheric conditions for the high temperature CO$_2$ capture. *Ind. Eng. Chem. Res.*, **2007**, *46*, 35-42.

[52] Adánez, J.; de Diego, L.F.; García-Labiano, F. Calcination of calcium acetate and calcium magnesium acetate: effect of the reacting atmosphere. *Fuel*, **1999**, *78*, 583-592.

[53] Lee, S.C.; Choi, B.Y.; Lee, T.J.;Ryu, C.K.; Ahn, Y.S.; Kim, J.C. CO$_2$ absorption and regeneration of alkali metal-based solid sorbents. *Catal. Today*, **2006**, *111*, 385-390.

[54] Lee, S.C.; Chae, H.J.; Lee, S.J.; Park, Y.H.; Ryu, C.K.; Yi, C.K.; Kim, J.C. Novel regenerable potassium-based dry sorbents for CO$_2$ capture at low temperatures. *J. Mol. Catal. B-Enzym.*, **2009**, *56*, 179-184.

[55] Martínez, I.; Grasa, G.; Murillo, R.; Arias, B.; Abanades, J.C. Kinetics of calcinations of partially carbonated particles in a Ca-looping system for CO$_2$ capture. *Energ.Fuel.*, **2012**, *26*, 1432-1440.

[56] Abanades, J.C.; Alonso, M.; Rodríguez, N.; González, B.; Grasa, G.; Murillo, R. Capturing CO₂ from combustion flue gases with a carbonation calcination loop. Experimental results and process development. *Energy Procedia*, **2009**, *1(1)*, 1147-1154.

[57] Rodríguez, N.; Alonso, M.; Grasa, G.; Abanades, J.C. Process for capturing CO₂ arising from the calcination of the CaCO₃ used in cement manufacture. *Environ. Sci. Technol.*, **2008**, *42 (18)*, 6980-6984.

[58] Charitos, A.; Rodríguez, N.; Hawthorne, C.; Alonso, M.; Zieba, M.; Arias, B.; Kopanakis, G.; Scheffknecht, G.; Abanades, J.C. Experimental validation of the calcium looping CO₂ capture process with two circulating fluidized bed carbonator reactors. *Ind. Eng. Chem. Res.*, **2011**, *50(16)*, 9685-9695.

[59] Alonso, M.; Rodríguez, N.; Grasa, G.; Abanades, J.C. Modelling of a fluidized bed carbonator reactor to capture CO₂ from a combustion flue gas. *Chem. Eng. Sci.*, **2009**, *64*, 883-891.

[60] Smith, A.R.; Klosek, J. A review of air separation technologies and their integration with energy conversion processes. *Fuel Process. Technol.*, **2001**, *70(2)*, 115-134.

[61] Darde, A.;Prabhakar, R.;Tranier, J.P.; Perrin, N. Air separation and flue gas compression and purification units for oxy-coal combustion systems. *Energy Procedia*, **2009**, *1(1)*, 527-534.

[62] Kakaras, E.; Koumanakos, A.; Doukelis, A.; Giannakopoulos, D.; Vorrias, I. Oxyfuel boiler design in a lignite-fired power plant. *Fuel*, **2007**, *86(14)*, 2144-2150.

[63] Romeo, L.M.; Abanades, J.C.; Escosa, J.M.; Paño, J.; Giménez, A.; Sánchez-Biezma, A.; Ballesteros, J.C. Oxyfuel carbonation/calcination cycle for low cost CO₂ capture in existing power plants. *Energ. Convers. Manage.*, **2008**, *49(10)*,2809-2814.

[64] Romeo, L.M.; Lara, Y.; González, A. Reducing energy penalties in carbon capture with organic Rankine cycles. *Appl. Therm. Eng.*, **2011**, *31*, 2928-2935.

[65] Romeo, L.M.; Lara, Y.; Lisbona, P.;Escosa, J.M. Optimizing make-up flow in a CO₂ capture system using CaO. *Chem. Eng. J.*, **2009**, *147*, 252-258.

[66] Metz, B.; Davidson, O.; de Connick, H.; Loos, M.; Meyer, L. (eds.) Special report on carbon dioxide capture and storage. *Technical report, Intergovernmental Panel on Climate Change*, **2005**.

[67] Alonso, M.; Rodríguez, N.; González, B.; Grasa, G.; Murillo, R.; Abanades, J.C. Carbon dioxide capture from combustion flue gases with a calcium oxide chemical loop. Experimental results and process development. *Int. J. Greenh. Gas Con.*, **2010**, *4*, 167-173.

[68] Charitos, A.; Hawthorne, C.; Bidwe, A.R.; He, L.; Scheffknecht, G. Design of a dual fluidised bed system for the post-combustion removal of CO₂ using CaO. Part II. Scaled cold model investigation. *9ᵗʰ International Conference on Circulating Fluidized Beds*, Hamburg, Germany, **2008**.

[69] Hawthorne, C.; Charitos, A.; Perez-Pulido, C.A.; Bing, Z.; Scheffknecht, G. Design of a dual fluidised bed system for the post-combustion removal of CO₂ using CaO. Part I. CFB carbonator reactor model.*9ᵗʰ International Conference on Circulating Fluidized Beds*, Hamburg, Germany, **2008**.

[70] Charitos, A.; Hawthorne, C.; Bidwe, A.R.; Sivalingam, S.; Schuster, A.; Splietho, H.; Scheffknecht, G. Parametric investigation of the calcium looping process for CO₂ capture in a 10 kW_th dual fluidized bed. *Int. J. Greenh. Gas Con.*, **2010**, *4*, 776-784.

[71] Pfeifer, C.; Puchner, B.; Hofbauer, H. *In-situ* CO₂-absorption in a dual fluidized bed biomass steam gasifier to produce a hydrogen rich syngas. *Int. J. Chem. React. Eng.*, **2007**, *5*, 15.

[72] Soukup, G.; Pfeifer, C.;Kreuzeder, A.; Hofbauer, H. *In situ* CO$_2$ capture in a dual fluidized bed biomass steam gasifier - Red material and fuel variation. *Chem. Eng. Technol.*, **2009**, *32*, 348-354.

[73] Koppatz, S.; Pfeifer, C.; Rauch, R.; Hofbauer, H.; Marquard-Moellenstedt, T.; Specht, M. H$_2$ rich product gas by steam gasification of biomass with *in situ* CO$_2$ absorption in a dual fluidized bed system of 8 MW fuel input. *Fuel Process. Technol.*, **2009**, *90*, 914-921.

[74] Hughes, R.W.; Lu, D.Y.; Anthony, E.J.; Macchi, A.Design, process simulation and construction of an atmospheric dual fluidized bed combustion system for *in situ* CO$_2$ capture using high-temperature sorbents. *Fuel Process. Technol.*, **2005**, *86*, 1523-1531.

[75] Lu, D.Y.; Hughes, R.W.; Anthony, E.J. Ca-based sorbent looping combustion for CO$_2$ capture in pilot-scale dual fluidized beds. *Fuel Process. Technol.*, **2008**, *89*, 1386-1395.

[76] Wang, J.; Manovic, V.; Wu, Y.; Anthony, E.J. A study on the activity of CaO-based sorbents for capturing CO$_2$ in clean energy processes. *Appl. Energ.*, **2010**, *87*, 1453-458.

[77] Fang, F.; Li, Z.; Cai, N. Continuous CO$_2$ capture from flue gases using a dual fluidized bed reactor with calcium-based sorbent. *Ind. Eng. Chem. Res.*, **2009**, *48*, 11140-11147.

[78] Dieter, H.;Bidwe, A.R.; Hawthorne, C.; Schuster, A. Design and construction of a 200 kW$_{th}$ calcium looping dual fluidized bed facility for CO$_2$ capture.*1st Meeting of the High Temperature Solid Looping Cycles Network*, Oviedo, **2009**.

[79] Hawthorne, C.; Dieter, H.; Bidwe, A.; Schuster, A.; Scheffknecht, G.; Unterberger, S.; Käss, M. CO$_2$ capture with CaO in a 200kW$_{th}$ dual fluidized bed pilot plant. *Energy Procedia*, **2011**, *4*, 441-448.

CHAPTER 6

Liquid-Gas Contactors Material Properties for CO_2 Capture in Absorption Column

Araceli Salazar[1], Rosa-Hilda Chavez[1,*] and Javier de J. Guadarrama[2]

[1]Instituto Nacional de Investigaciones Nucleares Carretera Mexico-Toluca S/N, La Marquesa, Ocoyoacac, 52750, Mexico and [2]Instituto Tecnológico de Toluca, Av. Instituto Tecnológico de Toluca S/N, Metepec, 52140, Mexico

Abstract: The purpose of this chapter is to discuss some structural characteristics and properties of three regular packing materials: metallic, polymeric and ceramic; in order to select the best one to capture CO_2 in an absorption column. The study was conducted by making the following tests: geometric physical properties such as wetted area and porosity; mechanical properties like, stress, hardness, modulus and compression resistance, structure and microstructure morphologic, chemical composition, rate of corrosion in electrochemical cell in medium of 1N of H_2SO_4 and Monoethanolamine (MEA) at 30% in aqueous solution by using standard procedures of the American Society of Testing Materials (ASTM) and own developed procedures for the equipment used. The structures of materials were also evaluated by X-ray diffraction and the surface of the material by scanning electron microscopy. It was concluded that metallic material is suitable for CO_2 gas treatment because it was presented with the lower etching 1N of H_2SO_4, and MEA at 30% in aqueous solution and giving the most absorption of CO_2.

Keywords: Liquid gas contactors, packing materials, material properties, subatomic structure, mechanical properties, tension test, hardness, tensile load, elongation, stress-strain behavior, poisson's rate, metallic bonding, austenitic stainless steel, polymers, ceramic, thermoplastic material, thermostable material, spectroscopy X-ray difraction, chemical resistance, scanning electron Microscope, absorption, morphology, porosity, separation efficiency.

1. INTRODUCTION

It is very common to use packed columns in sour gases treatment, which are also

*Corresponding author Rosa-Hilda Chavez: National Institute of Nuclear Research, Management of Environmental Science, Nuclear Center "Dr. Nabor Carrillo Flores", México-Toluca Road, La Marquesa, Zip Code 52750, Ocoyoacac Estado de Mexico, Mexico; Tel: +52 5553-297200 Ext 12654; E-mail: rosahilda.chavez@inin.gob.mx

known as packed towers. Packed column is used for the continuous contact of the liquid and gas flow, both flowing countercurrent inside the column filled with packing material as shown in Fig. **1**. The liquid is distributed over and trickles down through the packed bed, so that a large surface exposed to contact with the gas, hence the importance of characterizing materials contactors by their physical and chemical properties in order to determine the best regular packing for the treatment of carbon dioxide, such as:

1. Provide a large interfacial area between liquid and gas. The surface of the packing per packed bed should be large.

2. Possess large void fraction in the packed bed. Packing should allow passage of large volumes of fluid through small cross sections of the column, without, it means the gas pressure drop should be low.

3. The material must be chemically inert with respect to fluids being processed.

4. To be structurally strong to allow easy handling and installation, having low price.

5. There are primarily two types of packing: random and regular. This chapter treats about the study of three regular packing.

Figure 1: Packed absorption column with ceramic regular packing material.

1.1. Regular Packing Materials

Regular packing offers advantages such as lower pressure drop and grater liquid-gas contacted area, for higher gas flow (Fig. **2a, b, c**), but more expensive than random packing.

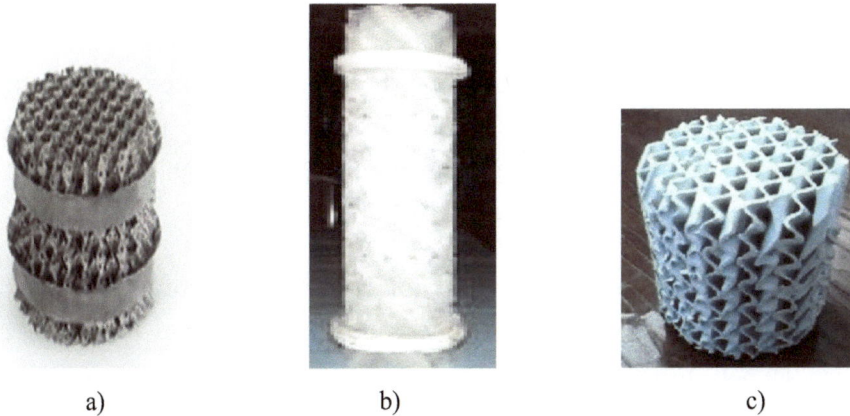

a)   b)   c)

Figure 2: Sulzer regular packing materials. a) metallic, b) polymeric, c) ceramic.

1.2. Material Properties

An engineer can properly select the material that meets the functional requirements of a part or device, considering their behavior and needs of economic production [1-3], for that is necessary to have a broad knowledge about:

a) The properties of the materials available.

b) The possibilities of manufacture, including the adaptability of various materials to the process, this means, the important properties affecting the process.

c) The effects of different processes on the properties of the materials.

Making a satisfactory choice is necessary to consider the above settings together, which means that not just select the cheapest material that meets some of the desired features, because it could be very expensive to process. Usually process changes the properties of materials; certain mechanical, physical and metallurgical changes, sometimes are beneficial and others are detrimental [4, 5].

The properties of the materials can be divided into the following four groups: a) Physical, b) Chemical, c) Mechanical and d) Technological [6, 7].

The physical properties include color, density, melting point, thermal expansion, electric conductivity among others [8].

Among chemical properties corrosion resistance plays an important role in the selection of materials and generally includes resistance to chemical or electrochemical attack [9]. The corrosion resistance is also important because it influences the formation of surface films affecting friction and lubrication as well as thermal and electrical conductivity [10].

Mechanical properties generally include material reactions to mechanical load. In most cases the main responsibility of the engineer in material selection is related to the mechanical properties, because to evaluate the performance of materials in terms of the desired functions, needs to know how materials react to design loads [11, 12].

The manufacturing and technological properties of a material describe the adaptability of a material to a particular process; these are very complex and usually are designed to test different methods to evaluate different characteristics.

The study of the properties of materials is of vital importance to relate them with their structure, based on the Science of Materials which is a multidisciplinary science and fundamental knowledge about the physical and macroscopic properties and their application in some areas of the science and technology [13, 14]. Strictly speaking, "materials science" involves the investigation of the relationships between the structures and properties of materials [14]. Structure is at this point a nebulous term that deserves some explanation. The structure of a material usually relates to the arrangement of its internal components. Subatomic structure involves electrons within the individual atoms and interactions with their nuclei. On an atomic level, structure encompasses the organization of atoms or molecules relative one to another. The next larger structural realm, which contains large groups of atoms that are normally agglomerated together, is termed "microscopic," which means that is subject to direct observation using some type of microscope. Finally, structural elements that may be viewed with the naked eye are termed "macroscopic." The notion of "property" deserves elaboration. While in service use, all materials are exposed to external stimuli that evoke some type of response. For example, a specimen subjected to forces will experience deformation, or a polished metal surface will reflect light [15-17]. A property is a material trait in terms of the kind and magnitude of response to a specific imposed stimulus. Generally, definitions of properties are made independent of material shape and size.

Virtually all important properties of solid materials may be grouped into six different categories: mechanical, electrical, thermal, magnetic, optical and deteriorative. For each one there is a characteristic type of stimulus capable of provoking different responses. Solid materials have been conveniently grouped into three basic classifications: metals, ceramics, and polymers [15, 16]. Some of the important properties of solid materials depend on geometrical atomic arrangements, and also the interactions that exist among constituent atoms or molecules. An understanding of many of the physical properties of materials is predicated on knowledge of the interatomic forces that bind the atoms together [17, 18]. The principles of atomic bonding are best illustrated by considering the interaction between two isolated atoms as they are brought into close proximity from an infinite separation [19]. At large distances, the interactions are negligible, but as the atoms approach, each exerts forces one on the other. These forces are of two types, attractive and repulsive, and the magnitude of each is a function of the separation or interatomic distance. The origin of an attractive force depends on the particular type of bonding that exists between the two atoms [19].

1.3. Mechanical Properties

Materials and metallurgical engineers are concerned with producing and fabricating materials to meet service requirements as predicted by these stress analyses. This necessarily involves an understanding of the relationships between the microstructure (*i.e.*, internal features) of materials and their mechanical properties.

Materials are frequently chosen for structural applications because they have desirable combinations of mechanical characteristics. The present discussion is confined primarily to the mechanical behavior of metals; polymers and ceramics are treated separately because they are, to a large degree, mechanically dissimilar to metals. This Chapter discusses the stress-strain behavior of three materials: metals, polymers and ceramic and the related mechanical properties.

If a load is static or changes relatively slowly with time and is applied uniformly over a cross section or surface of a member, the mechanical behavior may be ascertained by a simple stress-strain test; these are most commonly conducted for metals at room temperature. There are three principal ways in which a load may be applied: namely, tension, compression, and shear.

The interrelationship between processing, structure, properties, and performance is as depicted in the schematic illustration shown in Fig. **3**.

Figure 3: The four components of the discipline of materials science and engineering and their interrelationship.

Many times, a materials problem is concerned to select the right material from those available. There are several criteria on which the final decision is normally based [20]. First of all, the in-service conditions must be characterized, for these it is necessary to know the properties required for a specific service of the material [21]. On rare occasions the material possesses the maximum or ideal combination of properties, thus, it may be necessary to trade off one characteristic for the other [22, 23]. The classic example involves strength and ductility; normally, a material having a high strength will have only a limited ductility. In such cases, a reasonable compromise between two or more properties may be necessary. A second selection consideration is any deterioration of material properties that may occur during service operation. For example, significant reductions in mechanical strength may result from exposure to elevated temperatures or corrosive environments, so it is necessary to look for the adequate material. Finally, probably the overriding consideration is the economics.

1.4. Tension Test

One of the most common mechanical evaluations is stress-strain test [22]. As will be seen, the tension test can be used to ascertain several mechanical properties that are important in design. A specimen is deformed, usually until fracture, with a gradual increasing tensile load that is applied uniaxial along with the axis of a specimen with Universal Machine (Fig. **4**). A standard tensile specimen, cross section is normally circular, but rectangular specimens are also used. This "dogbone" specimen configuration was chosen so that, during testing, deformation is confined to the narrow center region (which has a uniform cross section along its length), and, also, to reduce the likelihood of fracture at the ends of the specimen.

The standard diameter is approximately 12.8 mm, whereas the reduced section length should be at least four times this diameter; 60 mm. It is common that gauge length is used in ductility computations; the standard value is 50 mm. The specimen is mounted by its ends into the holding grips of the testing apparatus. The tensile testing machine is designed to elongate the specimen at a constant rate, and to continuously and simultaneously measure the instantaneous applied load (with a load cell) and the resulting elongations (using an extensometer). A stress-strain test typically takes several minutes to perform; that is, the test specimen is permanently deformed until it is fractured.

The output of such a tensile test is recorded (usually on a computer) as load or force *versus* elongation. These load-deformation characteristics are dependent on the specimen size. For example, it will require twice the load to produce the same elongation if the cross-sectional area of the specimen is doubled. To minimize these geometrical factors, load and elongation are normalized to the respective parameters of engineering stress and engineering strain. Engineering stress is defined by the relationship: $\sigma = F/A$.

1.5. Stress-Strain Behavior

The degree to which a structure deforms or strains depends on the magnitude of an imposed stress. For most metals that are stressed in tension and at relatively low levels, stress and strain are proportional to each other through the relationship: $\sigma = E\varepsilon$. This is known as Hooke's law, and the constant of proportionality E (GPa) is the modulus of elasticity, or Young's modulus. For most typical metals the magnitude of this modulus ranges between 45 GPa for magnesium, and 407 GPa, for tungsten. Elasticity modulus values for several metals at room temperature are presented in Table **1**.

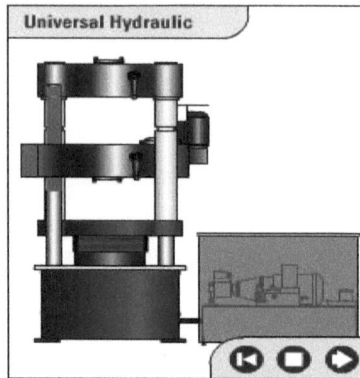

Figure 4: Universal machine to tension and compression test.

Table 1: Room-Temperature Elastic and Shear Modulus, and Poisson's Ratio for various Metal Alloys

Metal Alloy	Modulus of Elasticity		Shear Modulus		Poisson´s Rate
	GPa	10^6 psi	GPa	10^6 psi	
Aluminium	69	10	25	3.6	0.33
Brass	97	14	37	5.4	0.34
Copper	110	16	46	6.7	0.34
Magnesium	45	6.5	17	2.5	0.29
Nickel	207	30	76	11	0.31
Stell	207	30	83	12	0.34
Titanium	107	15.5	45	6.5	0.30
Tungsten	407	59	160	23.2	0.28

The mechanical behavior of materials is decisive for its use. Within the set of mechanical properties hardness is a characteristic relatively easy to determine and it provides valuable information, causing minimum damage to the specimen or sample. This practice focuses on testing the hardness, and raises the following objectives.

The instrument used to measure the hardness is named Durometer. Theoretical hardness is a measure of the material resistance to plastic deformation, such as the resistance to be scratched by other hard material, or the resistance of a material to be penetrated with an indenter which generates a trace on surface of material when is applied a controlled load (See Table **2**). Today hardness tests consist in measuring the depth or size of the footprint generated by an indenter. This test is an alternative in the nondestructive essay in the practice because the trace generated is minimal and its versatility is good, and because the hardness can estimate other properties as the tensile strength, which is more costly by the direct terms of equipment and sample preparation, depending on the indenter geometry and applied loads. In the following table collective test techniques common in hardness, materials are shown: The choice of one or other technique will be based on the type of material under study and its hardness. Hardness tests have a number of advantages that make its practice a usual mechanical characterization of materials: Its simplicity and low cost, requiring no special sample preparation.

Table 2: Types of Hardness and the different loads and penetrators to use in the test

Test of Hardness	Penetrator	Load	Formulae
Brinell	Ball of 10 mm of diameter of steel of tungsten carbide	29430 N	$HB = \dfrac{2P}{\pi D[D-\sqrt{D^2-d^2}]}$ P = Load, D= Diameter of penetrator, d = Diameter of indentation
Microhardness HV	Square based pyramidal diamond, angle of 136°	9.81N, 98.1N	$HV = \dfrac{1,854\,P}{d^2}$ P = Load, d = Diameter of indentation
Microhardness HK	Rhombic based pyramidal diamond	9.81N	$HK = \dfrac{14,2P}{l^2}$
Rockwell Hardness	Diamond Spheroconical penetrator, angle of 120°	588.6N (Rockwell A), 1471.5N (Rockwell C)	Lecture of hardness in direct form of the durometer caratule.
Rockwell superficial	Balls of steel of 1/16, 1/8, ¼, ½ of diameter	147.15N, 294.3N, 441.45N	Lecture of hardness in direct form of the durometer caratule.

1.6. Type of Bonding

To have an understanding of interatomic bonding in solids is that, in some instances, the type of bond allows us to explain the material properties.

An understanding of many physical properties of materials is predicated on the knowledge of the interatomic forces that bind the atoms together. Perhaps the principles of atomic bonding are best illustrated by considering the interaction between two isolated atoms as they are brought into close proximity from an infinite separation. At large distances, the interactions are negligible, but as the atoms approach, each exerts forces on the other. These forces are of two types, attractive and repulsive, and the magnitude of each is a function of the separation or interatomic distance. The origin of an attractive force FA depends on the particular type of bonding that exists between the two atoms. The magnitude of the attractive force varies with the distance. Ultimately, the outer electron shells of the two atoms begin to overlap, and a strong repulsive force FR comes into play. The net force FN between the two atoms is just the sum of both attractive and repulsive components; that is, $FN = FA + FR$.

Metallic bonding the final primary bonding type is found in metals and their alloys.

A relatively simple model has been proposed that very nearly approximates the bonding scheme. Metallic materials have one, two, three valence electrons. They may be thought of as belonging to the metal as a whole, or forming an "electron sea" or an "electron cloud". The remaining nonvalence electrons and atomic nuclei form that are called *cores ion,* which possess a net positive charge equal in magnitude to the total valence electron charge per atom, is an explication of metallic bonding. The free electrons shield the positively charged ion cores from mutually repulsive electrostatic forces, which they would otherwise exert upon one another; consequently the metallic bond is no directional in character. In addition, these free electrons act as a "glue" to hold the ion cores together [5].

In **covalent bonding** the stable electron configurations are assumed by the sharing of electrons between adjacent atoms. Two atoms that are covalently bonded contributes each one at least with one electron to the bond, and the shared electrons may be considered o belong to both atoms [6].

Ionic bonding is perhaps the easiest to describe and visualize. It is always found in compounds that are composed of both metallic and nonmetallic elements,

elements that are situated at the horizontal extremities of the periodic table. Atoms of a metallic element easily give up their valence electrons to the nonmetallic atoms. In the process all the atoms acquire stable or inert gas configurations and, in addition, an electrical charge; that is, they become ions. Sodium chloride (NaCl) is the classic ionic material [7]. A sodium atom can assume the electron structure of neon (and a net single positive charge) by a transfer of its one valence 3s electron to a chlorine atom. After such a transfer, the chlorine ion has a net negative charge and an electron configuration identical to that of argon. In sodium chloride, all the sodium and chlorine exist as ions. The study of packing materials was of great importance, because three types of materials were studied to determine the physical and chemical properties, next are general characteristics of the three types of materials studied.

1.6.1. Austenitic Stainless Steel

The term austenitic stainless steel refers to a class of iron based alloy containing sufficient chromium to prevent general corrosion in atmospheric exposure and sufficient austenizing elements (mainly nickel) to maintain an austenitic structure in the most common form, the alloys contain 18% Cr and 8% Ni. They are alloys of choice for many of the applications which require a higher level of corrosion resistance than carbon steel.

1.6.2. Polymers

The polymers are materials formed by the bend of many units, are chains of many macromolecules with molecular weight of 10000 or more of 1000000 of g/mol, they are classified as thermoplastic and thermostables. In this study, the polymeric contactor is of polipropilene which is a thermoplastic.

1.6.3. Thermoplastic

Thermoplastics are materials which become soften when they are heated to specific temperature and regain their original properties when they are cooled. And additional description might define as being linear polymers with little cross-linking. Crystallinity can be very high in certain polymers. This physical property relates with many of thermoplastics to the metals in mode of failure.

1.6.4. Ceramic

Ceramics, as a generic category, includes nearly everything that is primarily inorganic and nonmetallic, plastic or elastomeric [8]. Ceramics are compounds between metallic and nonmetallic elements, they are most frequently oxides carbides and nitrides, some of the common ceramics are: silica, aluminium oxide, nitride silicon. With regards to mechanical behavior ceramic materials are relatively stiff and strong and very hard. The contactor of ceramic in study resulted to be of aluminium silicate.

1.7. Methodology

The Methodology was conducted as follows:

1.7.1. Metrology of Length

The measurement was made using vernier and optical comparator Mitutoyo brand. The diameter of the packing, height, angle, the specific area was measured.

1.7.2. Mechanical Testing

1.7.2.1. Hardness Test

The hardness of the three materials was performed according to the Standard ASTM E384-84 Test Method for Microhardness.

1.7.2.2. Tensile Strength and Modulus of Elasticity

This trial was conducted according to the Standard E8-87 Test Method of Tension Testing of Metallic Materials.

1.7.2.3. Compression Strength

It was performed according to the Standard E9-87 compression Test of Metallic Materials.

1.7.3. Physico-Chemical Properties

1.7.3.1. Density

The density was determined by weighing a sample of material in the air and water, using Archimedes principle to determine the volume. The scale used was analytical accuracy of one ten thousandth of a gram, OHAUS brand.

1.7.3.2. Melting Point

The melting point of the materials was determined with a type K thermocouple connected to a multimeter.

1.7.3.3. Chemical Composition

The chemical composition was determined in a simple way with a scanning electron microscope JEOL JSM-59 000 brand LV, coupled with an energy dispersive spectroscopy.

1.7.3.4. Spectroscopy X-Ray Diffraction

The study to determine the structure of materials was made with a Diffractometer SIEMENS D5000 with 35 kV and 25 mA.

1.7.3.5. Electrochemical Test

Corrosion resistance with 1N H_2SO_4, and MEA (monoethanolamine 30 weight %)

It was analyzed according to ASTM G5-91 Standard Practice for Conducting Potentiodynamic Polarization Resistance Measurements.

1.7.3.6. Chemical Resistance with 1N H_2SO_4 (840h, and MEA (5540h)

The study was in accordance with ASTM G31-2004 Corrosion Testing Method gravimetric.

1.7.3.7. Scanning Electron Microscopy

The study of the surface was made with a Scanning Electron Microscope JEOL JSM-5900 brand LV before and after contact with H_2SO_4, and the aqueous solution of MEA 30 weight % according to ASTM E 986-86 Scanning Performance Characterization Electron Microscope to observe the morphology of the contactors materials before and after the electrochemical test.

1.7.3.8. Absorption of Carbon Dioxide into the Absorption Column

Packing materials: metallic, polymeric and ceramic were tested functionally within an absorption column, supplying 20 m^3/h of gas with composition of CO_2 weight % and MEA 30 weight %.

Results

The results presented for each regular packing material are as follows:

Metallic regular packing material: According to the results (Tables **3**, **4**) and the chemical composition (Table **5**) presented metallic contactor alloying elements Cr, Ni, Mo, which influence the material properties are as follows: Cr: This element increases tensile strength, yield strength, mechanical strength, hot hardness, hardenability, fatigue resistance, wear resistance, toughness, heat resistance, resistance to corrosion, which is the essential component in stainless steels. The element Molybdenum increases the tensile strength, hot strength, yield strength, fatigue resistance, hardness, wear resistance, toughness, hardenability, is a potent stabilizer of the carbides. The Nickel: Increases tensile strength, yield strength, mechanical strength, hot-hardenability, density, elongation, corrosion resistance and the risk of fragility hot. The Silicon increases the tensile strength, yield strength, hardness, strength, hot corrosion resistance, in short these alloying elements cause the material to resist oxidation, the microstructure corresponds to an austenitic steel, their structure is face-centered cubic (see diffractogram 1). This structure has the following characteristics: the number of atoms contained in the unit cell is 4: 1/2x6 atoms in the center of the faces and vertices 1/8x8 atoms with a coordination number of 12.

With regard to mechanical properties, the material of medium hardness, morphology before contact with 1N H_2SO_4 was the most rugged of the three (see Figs. **5**, **6**, **7**, **8**), and was the most resistance mechanically (Table **4**), chemically resistant (Tables **6** and **7**), the most efficient in the process of absorption (Table **8**, Fig. **5**).

Polymeric regular packing material: The results of melting point, density and elemental composition of the polymeric material correspond to the polypropylene (Tables **3** and **5**), it has monoclinic crystal structure, surface morphology is observed smoother than the other two materials studied (see Figs. **9, 10, 11**).

Ceramic regular packing material: This material presents elemental chemical composition of silicon, aluminum, oxygen, corresponding to an aluminosilicate (Table **5**), crystal structure of two phases for quartz SiO_2 hexagonal and orthorhombic for mullite (Al_2O_3) (see diffractogram 3), its denotes higher porosity surface morphology of the three materials studied (see Figs. **12, 13** and **14**).

Table 3: Physical properties of regular packing materials

Characteristics	Metallic Regular	Polymeric Regular	Ceramic Regular
Diameter in cm	6.5cm	8.5cm	9.3
High in cm	17cm	32cm	10.3
Angle	35°	45°	45°
Specific Area (m^2/m^3)	498 m^2/m^3	264 m^2/m^3	180 m^2/m^3
Void fraction (m^3/m^3)	0.90 m^3m^3	0. 90 m^3/m^3	0.91 m^3/m^3
Density (g/cm^3)	7.9 g/cm^3	0.91 g/cm^3	2.43 g/cm^3
Flow canal	5	4	7
Weight	150 g	180 g	465 g
Melting point	1400°C	170°C	Over of 1400°C

Table 4: Mechanical properties of regular packing materials

Mechanical Property	Metallic Regular	Polymeric Regular	Ceramic Regular
Tension Test	841 MPa	35 MPa	90 MPa
Compression test	130 kg/cm^2	100 kg/cm^2	3200 kg/cm^2
Knoop Hardness	190 HK	20 HK	900 HK
Elasticity Modulus	200 GPa	1300 N/mm^2,1.2 GPa	145 GPa

Table 5: Chemical elemental composition of regular packing materials

Elemental Composition	Metallic Regular (%)	Polymeric Regular (%)	Ceramic Regular (%)
Al	0	0	16.6
Mn	0.7	0	0
S	0.02	0	0
C	0.02	*87	0
P	0.01	0	0
Si	0.20	0	22.9
Cr	18	0	0
Ni	13	0	0
Mo	2.2	0	0
K	0	0	1.7
O	0	0	58.8
Fe	Balance: 34.15	0	0

* The hydrogen balance of 13% is not reported by this technique.

Table 6: Corrosion rate of the contactor materials, in presence of H$_2$SO$_4$ 1N, by electrochemical and gravimetric method

Contactor Material	Rate of Corrosion at H$_2$SO$_4$ 1N (Electrochemical Method)	840h at H$_2$SO$_4$1N (Gravimetric Method)
Metallic	7.8x10^{-4} mpy	3.12x10^{-1} mpy
Polymeric	2.82x10^{-4} mpy	1.13x10^{-1} mpy
Ceramic	1.03x10^{-1} mpy	1.58x10^{-2} mpy

Table 7: Corrosion rate of the contactors materials, with aqueous solution of MEA, by electrochemical and gravimetric method

Material	Rate of Corrosion at Aqueous Solution of MEA 30% (Electrochemical Method)	5540h at Aqueous Solution of MEA 30% (Gravimetric Method)
Metallic	6.42x10^{-2} mpy	6.19x10^{-1}mpy
Polymeric	1.48x10^{-1}mpy	2.01x10^{1}mpy
Ceramic	5.6x10^{-1}mpy	2.57mpy

Table 8: Separation efficiency of three regular packing materials, after 70 minutes working into packed absorption column

Time in Minutes	(%) Absorption of CO$_2$		
	Metallic Regular	Polymeric Regular	Ceramic Regular
1	95	93	84
5	93	92	80
10	90	88	74
20	89	84	70
30	77	76	68
40	74	64	62
50	67	52	50
60	66	46	44
70	55	35	32
Observations	Roughness surface, wich allowed more area of contact in the mass transference process.	Hidrofobic material. This caused more time to form the liquid film.	Porous material, hidrophilic, It is easy humidificated, for that offers minor area between the liquid and the gas during the absorption process.

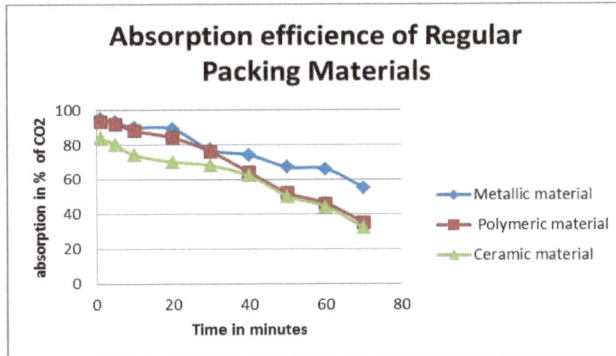

Figure 5: Absorption´s efficiency of CO_2 using regular packing materials: Metallic, polymeric and ceramic.

Stainless Steel

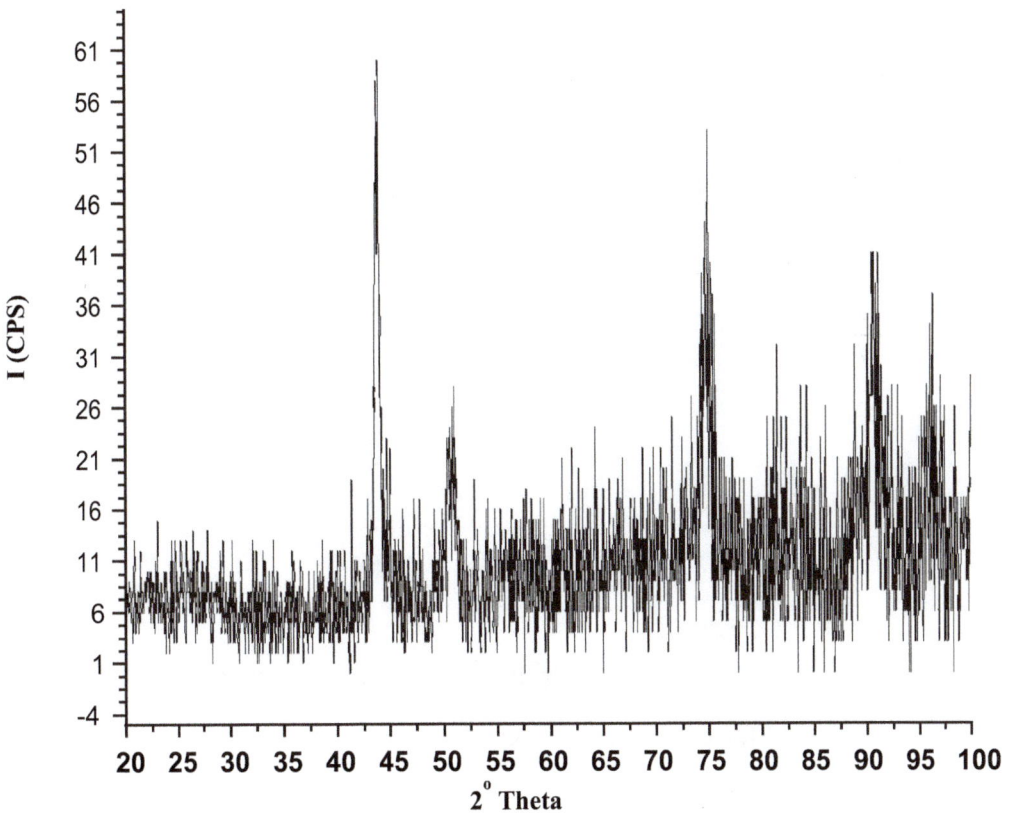

Difractogram 1. Metallic Regular Packing, it Shows Centered Cubic Structure Faces

This difractogram number 1 corresponds to the metallic material, in which overhangs four diffraction lines located at 43.58°, 50.79°, 74.69°, 90.68° and 95.60°, which coincide with the diffraction lines of austenite crystalline phase (JCPDS 033-0397).

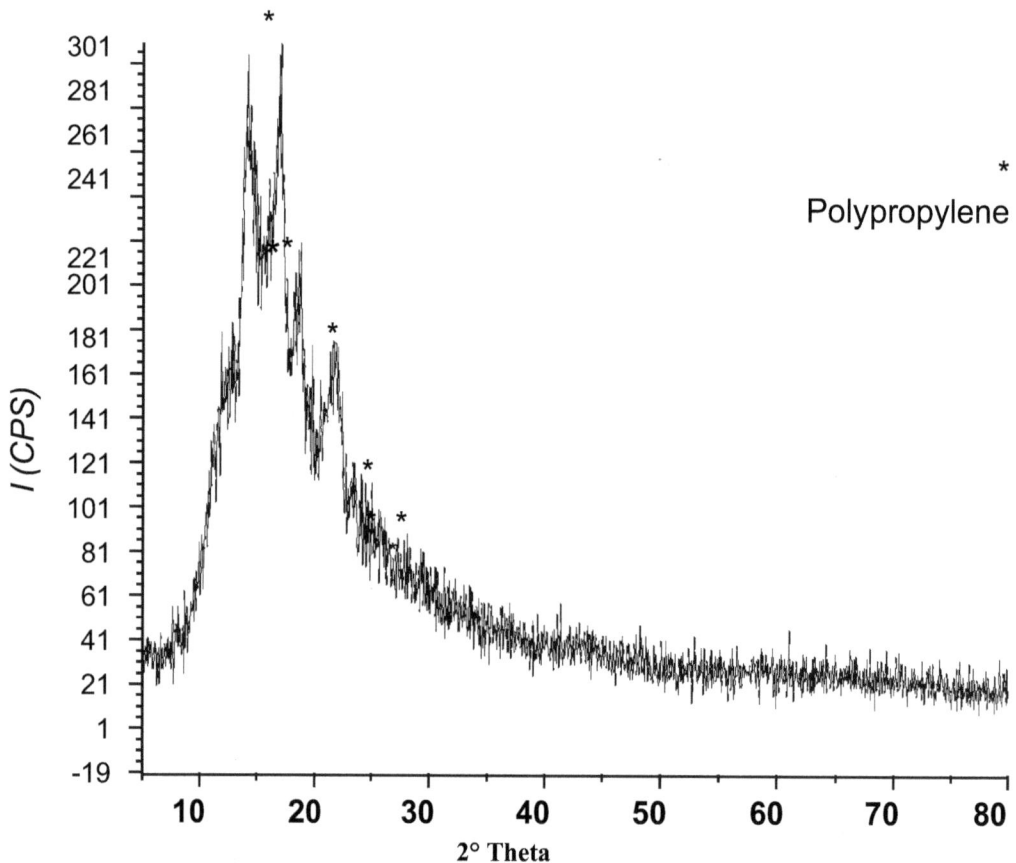

Difractogram 2. Polymeric Regular Packing, it Shows Monoclinic Structure

Analysis of X-ray diffraction performed on the polymeric material, number 2, revealed intense diffraction lines located at 14.08°, 17.08°, 18.49°, 21.78°, which coincide with the diffraction lines of a crystalline phase called polypropylene (JCPDS 054-1936). This pattern corresponds to the polypropylene monoclinic phase, which promotes increased toughness in the material.

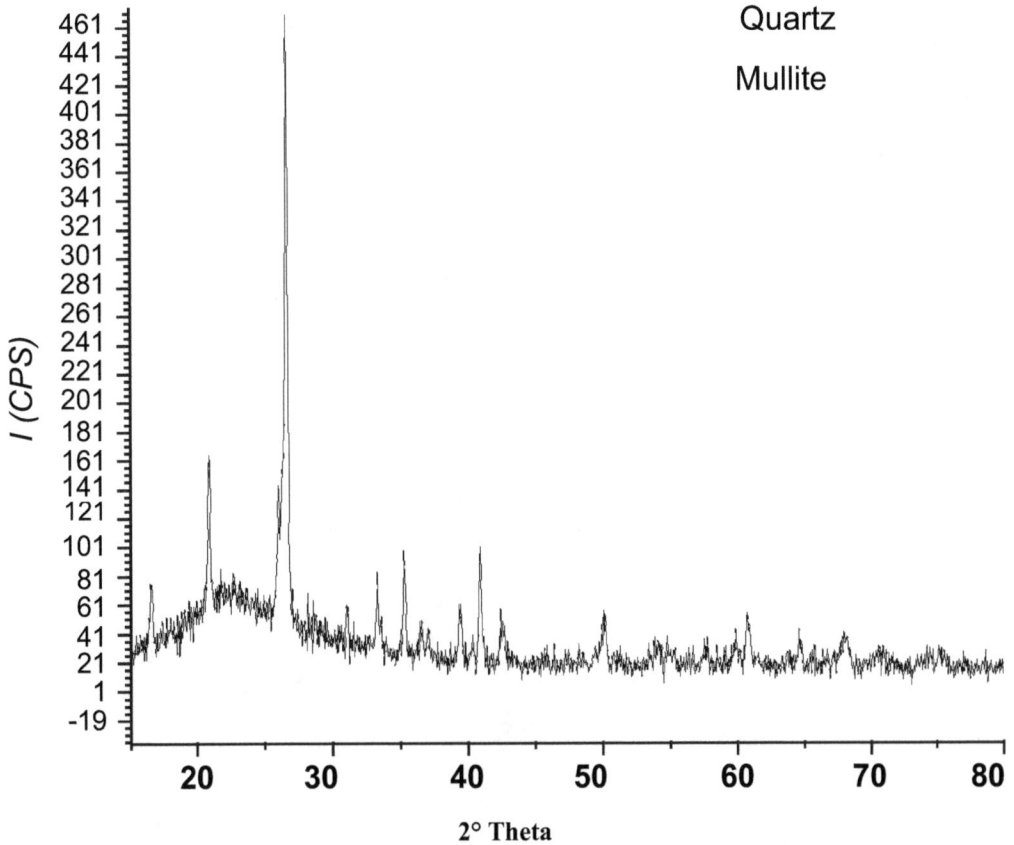

Difractogram 3. Ceramic Regular Packing, it Shows two Structures Characteristics (Quartz Hexagonal, and Mullite Orthorombic)

The analysis of X-ray diffraction of packing ceramic, showed several lines located at 2θ diffraction: $20.8°$, $26.6°$, $39.4°$, $45.8°$, $50.1°$, $59.9°$, $63.9°$, $57.7°$, $73.4°$, corresponding to the hexagonal phase quartz (JCPDS 70-3755), also showed the diffraction lines at 2θ located: $16.4°$, $26.0°$, $30.9°$, $33.1°$, $35.2°$, $40.8°$, $42.9°$, $60.7°$, are attributed to the mullite crystalline phase (JCPDS 74-4145).

The morphology of the regular packaging technique studied by scanning electron microscopy before and after they were in contact with the MEA solution 30 weight % and with sulfuric acid 1 N, it was observed the morphological changes in its surface which are showed in Fig. **6** to **14**.

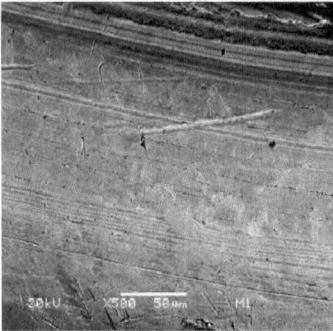

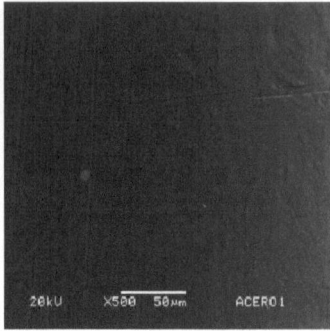

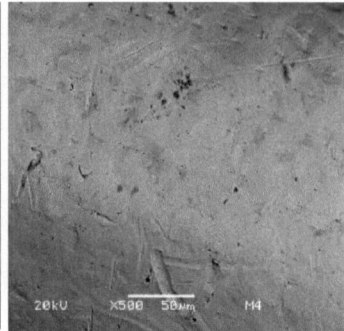

Figure 6: Metallic packing surface rouhgness before the essay with H$_2$SO$_4$ and MEA.

Figure 7: The surface metallic packing shows minus roughness after the essay with H$_2$SO$_4$.

Figure 8: Metallic packing surface with zones blacker by the Cr$_2$O$_3$, formed after the essay with MEA solution.

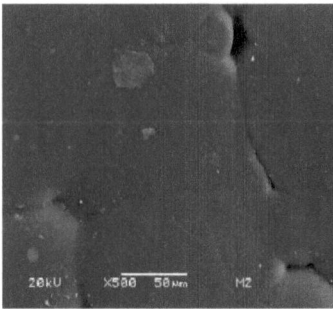

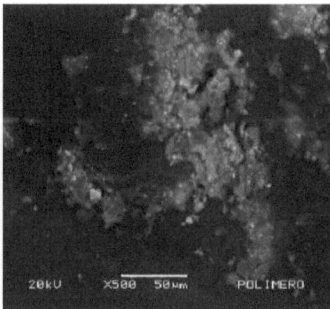

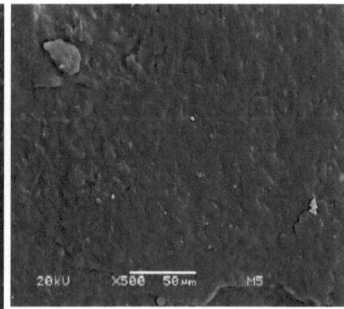

Figure 9: Flat surface of polymeric packing before the essay with H$_2$SO$_4$ and MEA.

Figure 10: Surface of polymeric packing after the essay with H$_2$SO$_4$. It shows a roughness zone.

Figure 11: Surface of polymeric packing after the essay with MEA. It shows all surface more roughness.

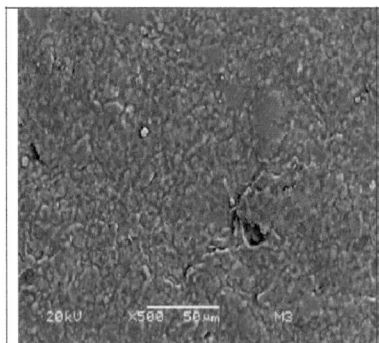

Figure 12: Porous surface of the ceramic packing before the essay with H$_2$SO$_4$ and MEA.

Figure 13: The surface shows attach by the action of H_2SO_4.

Figure 14: The ceramic surface shows more attack with the solution of MEA.

CONCLUSION

The study of regular packing materials allowed characterizing them in terms of their physical, mechanical and chemical properties as well as the resistance and efficiency of each one in the presence of sour gases in absorption column.

The metallic material for its mechanical properties and chemical composition resulted to be the most suitable for using in the absorption column with MEA due to form a surface layer of chromium oxide Cr_2O_3, which serves as a barrier to prevent further oxidation of the material. The efficiency for the treatment of carbon dioxide was 95% of CO_2 absorption.

The polymeric material proved to be one of the two most resistant in the presence of aqueous 1N sulfuric acid, and the second resistance in the presence of aqueous solution of MEA 30 weight %, making it a candidate for use in the treatment of SO_2. Regarding the treatment of CO_2, the absorption of the gas resulted in 93%, which is a percentage very attractive for use at large scale.

The ceramic material was the least resistant compared to the other two materials. However, for small-scale treatments can be used, as it gave an efficiency of 84% in the separation of carbon dioxide, non-negligible amount.

ANNEXES: INSTRUMENTAL TECHNIQUES USED TO EVALUATE THE REGULAR PACKING MATERIALS

1. ENERGY-DISPERSIVE X-RAY SPECTROSCOPY (EDS)

When the atoms in a material are ionized by a high-energy radiation they emit characteristic X-rays. EDS is an acronym describing a technique of X-ray spectroscopy that is based on the collection and energy dispersion of characteristic X-rays. An EDS system consists of a source of high-energy radiation, usually electrons; a sample; a solid state detector, usually made from lithium-drifted silicon, Si (Li); and signal processing electronics. EDS spectrometers are most frequently attached to electron column instruments. X-rays that enter the Si (Li) detector are converted into signals which can be processed by the electronics into an X-ray energy histogram. This X-ray spectrum consists of a series of peaks representative of the type and relative amount of each element in the sample. The number of counts in each peak may be further converted into elemental weight concentration either by comparison with standards or by standard calculations.

Range of Elements: Boron to Uranium

Destructive: No

Chemical Bonding Information: Not Readily Available

Quantification detection limits: Best with standards, although standard methods are widely used.

Accuracy: Nominally P 5%, relative, for concentrations > 5 % weight Detection limits: 100-200 ppm for isolated peaks in elements with Z >11.1-2% weight for low-Z and overlapped peaks

Lateral resolution: -5-1 pm for bulk samples; as small as 1 nm for thin samples in STEM.

Depth sampled: 0.02 to pm, depending on Z (atomic number) and keV Imaging/mapping: In SEM.

Sample requirements: Solids, powders, and composites; size limited only by the stage in SEM.

Main use: To add analytical capability to SEM.

2. X-RAY DIFFRACTION

In X-Ray Diffraction (XRD) a collimated beam of X-rays, with wavelength h-*0.5-2 Å,* is incident on a specimen and is diffracted by the crystalline phases in the specimen according to Bragg's law (h = 2dsinθ), where d is the spacing between atomic planes in the crystalline phase). The intensity of the diffracted X-rays is measured as a function of the diffraction angle 2θ and the specimen orientation. This diffraction pattern is used to identify the specimen crystalline phases and to measure its structural properties, including strain (which is measured with great accuracy), epitaxy, and the size and orientation of crystallites (small crystalline regions). XRD can also determine concentration profiles, film thicknesses, and atomic arrangements in amorphous materials and multilayers. It also can characterize defects. To obtain this structural and physical information from thin films, XRD instruments and techniques are designed to maximize the diffracted X-ray intensities, since the diffracting power of thin films is small.

Ideally, the more stable arrangement of the coordination polyhedron is a crystal that minimizes the energy per unit, in other words, one that:

1. Preserve electrical neutrality,

2. Meet the directionality and the discrete nature of all covalent,

3. Minimize the strong ion-ion repulsion,

4. Stack atoms as compactly as possible consistent with (1), (2) and (3).

Before discussing the actual three-dimensional diagrams consist of atoms in a crystal structure, it is useful that the diagrams are possible for identical points in space. Diagrams are called spatial networks and each crystal structure is based on one of those possible.

Bravais´s Space Network

A management network are three-dimensional spatial infinite points in which each have an environment identical to each other and are called grid points. These forms are arranged on one of 14 different lattices called Bravais lattices, and therefore the crystalline structure throughout atoms being in positions designated by the appropriate lattice. In each lattice point may be associated more than one atom, but each atom or group of atoms in a lattice point must have an atom or group of atoms identical, with the same orientation in any other point of the network to meet the definition of spatial network. Because a perfect crystal structure is a regular diagram of atoms arranged in a spatial network, the atomic arrangements can be described completely by specifying the atomic positions of the repeating unit of a spatial network. This unit is called the unit cell, and if specified atomic positions within the same unit cell are called the crystal structure, and its edges must be translational network (vectors connecting any two lattice points), the identical unit cells, a particular spatial grid and fill the space generated when the spatial network stack them face to face. A spatial network can have several different unit cells that meet the conditions described above, but by convention, the unit cells are chosen to present a simple geometry and contain only a few points in the network, such as the 14 Bravais lattices shown in Fig. **15**.

If the unit cell vectors (vectors parallel three coincident with the edges of the cell) are chosen such that the lattice points in a single cell occupy the vertices (or equivalently where only one lattice point per cell as can be seen in the slightly sliding space and a particular type is called primitive cell.

Only light elements of group IV have crystal structures in which all links that hold the glass are covalent. The joints are the result of superimposing sp3 hybrid orbitals and the resulting structure is cubic diamond. Graphite is a crystalline form of carbon in it, and the atoms are covalently bonded as flat hexagonal arrangements with weak secondary bonds between the planes.

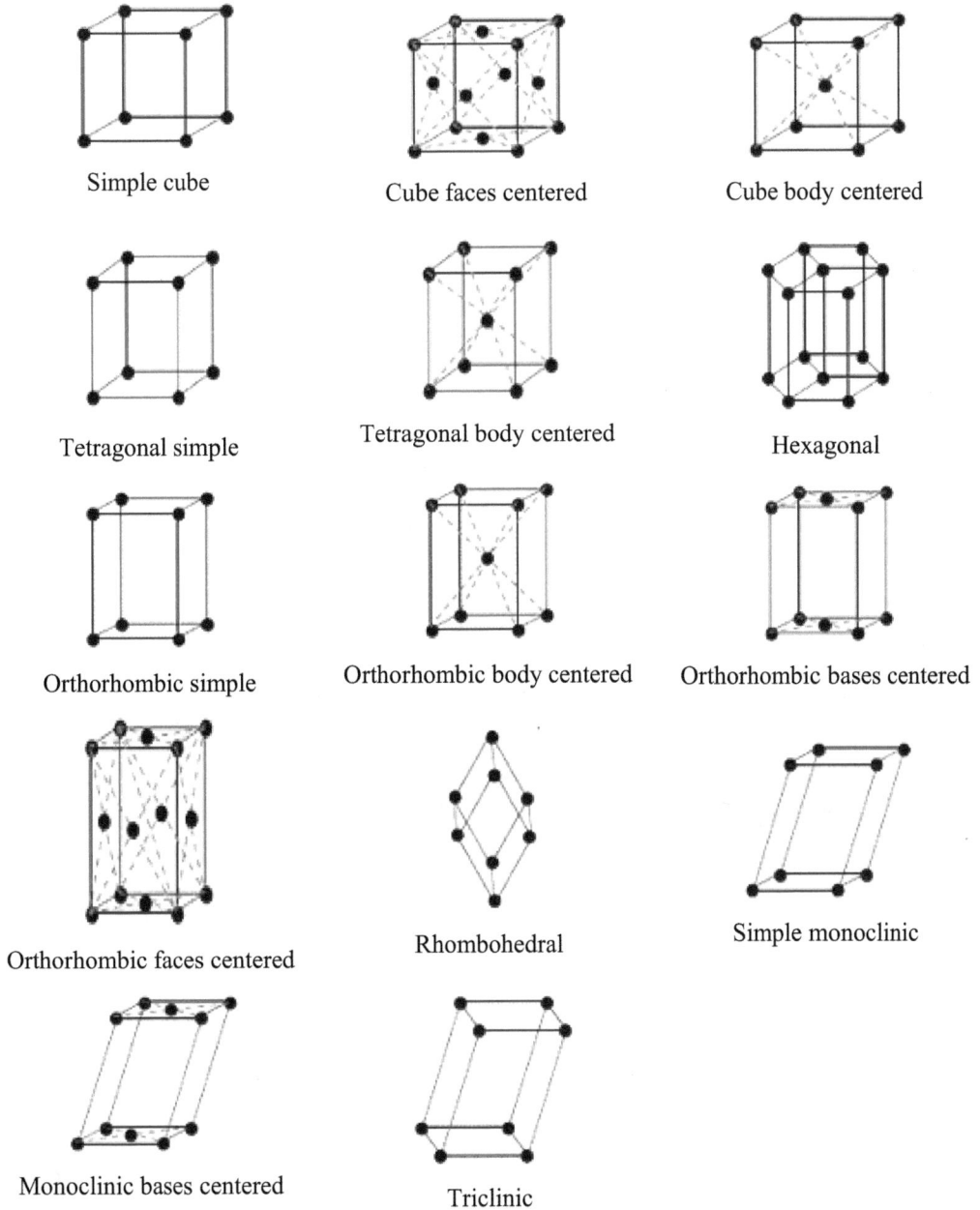

Figure 15: The 14 Bravais's Network.

Ionic crystals. Although very few are completely ionic crystals, many have a significant percentage of ionic and can be classified broadly as ionic crystals, so for example the case with NaCl, MgO, SiO$_2$, LiF. In purely ionic crystals, anionic polyhedral are stacked as to maintain electrical neutrality and minimize the binding energy per unit volume without introducing strong repulsion between ions of the same charge.

The X ray diffraction supplier the following advantages:

Range of Elements:

All, but not element specific, Low-Z elements may be difficult to detect.

Probing Depth:

Typically few pm, but it is highly material dependent; monolayer sensitivity with synchrotron radiation.

Detection Limits:

Material dependent, but -3% in a two phase mixture; with synchrotron radiation can be - 0.1%.

Destructive:

Not for most materials.

Depht Profiling:

Normally none, but this can be achieved.

Sample Requirements:

Greater than -0.5 cm, although smaller requires microfocus.

Lateral Resolution:

Normally none, 10 μm requires microfocus.

Main in Use:

Identification of crystalline phases, determination of strain, crystallite orientation and size, accurate determination of atomic arrangements.

Specialized Uses:

Defect imaging and characterization; atomic arrangements in amorphous materials and multilayers; concentration profiles with depth; film thickness measurements.

This is a technique used to determine the crystal structure of solid materials, since some of the most important properties of solid materials depend on the geometric arrangement of atoms and the interactions between the constituent atoms and molecules. X-rays are electromagnetic radiation of the order of 1 Angstrom, are located in the electromagnetic spectrum from UV rays and γ-rays, the interatomic distances between the solids are of the order of Angstroms, precisely the order of the wavelength of X-rays, allowing experiments on crystalline samples of DRX, as this radiation is the space between the crystalline planes of a solid grooves in a diffraction grating. Structural information on a solid material can be obtained by XRD to understand the physical and chemical properties of solid materials.

X-rays are produced by bombarding a metallic target, typically Mo or Cu, with electrons accelerated by an intense electric field 30-50 keV, from a filament. The electron beam incident on the metal target, X-rays are produced through 2 mechanisms: a) The incident electrons are slowed or stopped by the collision, and some of its lost energy is converted into X-rays, with wavelengths in a certain range, from a lower limit value, which is an upper limit for energy, which corresponds to the event in which all the kinetic energy of the electron becomes X-ray. In this situation the energy generated is called radiation Bremstralung. b) The incident electrons ionize K-shell electrons (1s) of the target atoms and X-rays are emitted as a result of relaxation of the system by filling these empty orbitals with electrons from higher energy layer L (2p) or M (3p), which gives rise to the characteristic emission lines of K α and K β, corresponding to the relaxations: LK α 1 , α 2 K and MK K β 1, β 2 respectively the wavelength of X-rays emitted is characteristic of each material and decreases with increasing atomic number, with 1.542 Angstroms of K α 0. 711 Angstroms of copper and the K α Molybdenum. So that the spectrum of X-ray emission has two components: the part of Bremstralung and spikes caused by electronic transitions within the atom.

3. SCANNING ELECTRON MICROSCOPY (SEM)

The Scanning Electron Microscope (SEM) is often the first analytical instrument used when a "quick look" at a material is required and the light microscope no longer provides adequate resolution.

In the SEM an electron beam is focused into a fine probe and subsequently raster scanned over a small rectangular area, as the beam interacts with the sample it creates various signals (secondary electrons, internal currents, photon emission, *etc.*), all of which can be appropriately detected.

These signals are highly localized to the area directly under the beam, by using these signals to modulate the brightness of a cathode ray tube, which is raster scanned in synchronism with the electron beam, an image is formed on the screen. This image is highly magnified and usually has the look a traditional microscopic image but with a much greater depth of field. With ancillary detectors, the instrument is capable of elemental analysis. Its main use is high magnification imaging and composition (elemental) mapping.

Scanning electron microscopy is an analytical technique where it is used electrons beam instead of light beam to form the image. Ithas a depth of field which allows to focus a large part of the sample and also produces highresolution images, which means that spatially close features in the sample can be examined with greater detail.

The sample preparation is relatively easy, since most SEMs only require the samples that are conductive, for which the sample is coated with a thin layer of carbon or a metal such as gold, to give conductive properties to the sample. After scanning is done with the accelerated electrons traveling through the barrel of the microscope, the detector measures the amount of mail sent resulting in the intensity of the sample area, being able to display three-dimensional figures, projected an image of TV or a digital image.

The resolution of the electron microscope is of the order of 3 to 20 nm, depending on the type of equipment. The scanning electron microscope was invented in 1931 by Ernst Ruska, allowing to approach the atomic world, it is possible get highresolution images from stone materials, metallic and organic. The light is replaced by an electron beam, the lenses by electromagnets and the samples are metallic conductive surface. This instrument allows observation and surface characterization of inorganic and organic, giving morphological information of the analyzed material. From the SEM results in different types of signals are generated from the sample and used to examine many of its features, with the technique of scanning electron microscopy morphological studies of microscopic areas of different materials can be performed, in addition to processing and analyzing the images obtained.

The procedure for observing a sample is as follows:

a) The sample is prepared by making driving so it is coated with gold.

b) Place the sample in the microscope sample holder.

c) Place the sample holder with the sample inside the microscope chamber.

d) It is empty.

e) Adjusting the working distance of 10 mm.

f) The sample is focused.

g) It is sweeping the area selected for the study.

h) Observe the sample required increases, according to the details you want to watch.

i) If required chemical analysis is chosen for this area.

j) Keeps the energy dispersion diagram on a CD.

ACKNOWLEDGEMENTS

Partial financial support of this work were provided by Consejo Nacional de Ciencia y Tecnología (CONACyT), project: CB-2007-01-82987 and EDOMEX-2009-C02-135728. All are greatly appreciated.

CONFLICT OF INTEREST

The authors confirm that this chapter contents have no conflict of interest.

REFERENCES

[1]   Baillie, C.; Vanasupa, L. *Navigating the Materials World,* Academic Press, San Diego, CA, 2003. Flinn, R.A. and P.K. Trojan, *Engineering Materials and Their Applications,* 4thed.; John Wiley & Sons, NY, **1994**.

[2]   Jacobs, J.A.; Ki Duff, T.F.*Engineering Materials Technology,* 5thed.; Prentice Hall PTR, Paramus, NJ, **2005**.

[3]   Mango, P.L.*The Principles of Materials Selection for Engineering Design,* Prentice Hall PTR, Paramus, NJ, **1999**.

[4]   Murray, G.T.*Introduction to Engineering Materials—Behavior, Properties, and Selection,* Marcel. Dekker, Inc., NY, **1993**.

[5]   Alting L. *Procesos para Ingeniería de Manufactura.*Alfaomega, **1990**.

[6]   White, M.A. *Properties of Materials,* Oxford University Press, NY, **1999**.

[7]   Ralls, K.M.; Courtney, T.H.; Wulff, J. *Introduction to Materials Science and Engineering,* John Wiley & Sons, NY, **1976**.

[8]   Callister, D.W. *Fundamentals of Materials Science and Engineering an introduction,* 7th ed.; Wiley, **1990**, pp. 178-224.

[9]   Callister, D.W. *Materials Science and Engineering an introduction,* 7th ed.; Wiley, **2001**, pp 132-150.

[10]   Schaffer, J.P.; Saxena, A.; Antolovich, S.D.; Sanders, T.H.J.; Warner, S.B. *The Science and Design of Engineering Materials,* 2nd ed.; WCB/McGraw-Hill, NY, **1999**.

[11]   Smith, W.F.; Hashemi, J. *Principles of Materials Science and Engineering,* 4th ed.; Paramus, NJ, **2006**.

[12]   Van Vlack, L.H.*Elements of Materials Science and Engineering,* 6th ed.; Addison-Wesley Longman, Boston, MA, **1989**.

[13]   Delly, J.G. *Photography Through The Microscope.* Eastman Kodak Company, Rochester, **1988**.

[14]   Duke P.J.; Michette, A.G. *Modern Microscopies* Plenum, NY, **1990**.

[15]   Pluta, M.*Advanced Light Microscopy.* Elsevier, Amsterdam, **1988.**

[16]   Callister D.W. *Material Science and Engineering.* Department of Metallurgical Engineering. The University of Utah with special contributions by David G. Rethwisch. The University of Iowa, 7th. ed.; John Wiley Sons, Inc. **2007**.

[17]   Mc Crone, W.C.; Delly. J.G. *The Particle* Atlas. Ann Arbor Science, Ann Arbor, **1973**; Vol. 1-4. Palenik, S., **1979**; Vol. 5. Brown, J.A.; Stewart, I.M., **1980**; Vol. 6.

[18]   Kehl, L. *The Principles of Metalhographic Laboratory fiatice.* Mc Graw-Hill, NY, **1949**.

[19]   Oelsner O. *Atlas of the Most Important Ore Mineral Parageneses Under the Microscope.* Pergamon, London,**1966**.

[20]   ASM Handbook Committee. *Metals Handbook,* Vol. 7: *Atlas of Microstructures.*American Society of Metals, Metals Park, **1972**

[21]   McMahon, C.J., Jr., *Structural Materials,* Merion Books, Philadelphia, **2004**.

[22]   ASTM Standards E 8 and E 8M, *Standard Test Methods for Tension Testing of Metallic Materials*. Annual Book of ASTM Standards. Vol, 03.01.Editorial Staff. 2000.

[23]   Benedetti-Pichler. *Identification of Materials.*Springer-Verlag, NY, **1964**.

CHAPTER 7

Rate-Based Models and Design of Packed Columns for Absorption of Carbon Dioxide

Andrés Emilio Hoyos-Barreto[1], Anja Müller[2], Felipe Bustamante[3,*] and Aída Luz Villa[3]

[1]Chemical Engineering Department, Universidad de Antioquia, Medellín, Colombia; [2]Laboratory of Fluid Separations Department of Biochemical and Chemical Engineering, TU Dortmund University, Emil-Figge-Straße 70, D-44227 Dortmund, Germany and [3]Environmental Catalysis Research Group, Chemical Engineering Department, University of Antioquia, Medellín, Colombia

Abstract: Comparison of rate-based models (RBM) and equilibrium models (EQM) for absorption of CO_2 in ammonia, along with details on selecting process equipment, are presented in the first part of the Chapter. Differences in temperature and concentration profiles along a 10-stage packed column were found with the two models. In EQM, liquid temperature matches the inlet gas temperature at the fourth stage and remains constant until the bottom of the column. Oppositely, three temperature profiles were obtained in RBM: interface, gas, and liquid streams; temperatures in the EQM were almost 10 °C higher than in the RBM. For intermediate stages, the EQM predicted higher mol fraction of CO_2 than RBM.

A non-equilibrium stage, rate-based model is developed in the second part of the chapter. Simultaneous mass and energy transfer through the gas-liquid interface, as well as the hydrodynamics of packed columns with the corresponding correlations to calculate pressure drop, liquid hold up and mass transfer coefficients, are included in the model which is based on absorption segments. Simulation of the absorption of CO_2, from a mixture with ammonia indicated more than 80% of absorption of CO_2; complete absorption of SO_2 and NO_2 present in the stream could also be accomplished. It was necessary to increase ammonia concentration in the lean solution and implement an ammonia washer in order to avoid ammonia slip. These results were used for sizing a packed column for the treatment of a gas stream with a composition similar to that expected from a wet cement kiln.

Keywords: Carbon dioxide, absorption, ammonia, design, rate-based models, equilibrium models, ASPEN, NO_x, SO_2.

---

*Corresponding author Felipe Bustamante: Environmental Catalysis Research Group, Chemical Engineering Department, Universidad de Antioquia, Colombia; Tel: (+57)-42198535/(+57)-42196609; E-mail: felipe.bustamante@udea.edu.co

1. INTRODUCTION

This chapter presents the basics of packed towers for gas absorption, and a comparison with tray columns and spray washers. It also provides a brief description of the different models used for simulating gas/liquid separation processes, as well as a comparison between equilibrium and non-equilibrium models (rate-based models).

Then, the structure of rate-based models is introduced and a rate-based model is developed and used for the simulation of the simultaneous absorption of CO_2, NO_2 and SO_2 in a packed column with high efficiency materials (third generation packing). Finally, the simulation results are used for sizing an unit for the treatment of an actual exhaust stream in a wet cement plant. The basis of this chapter is the PhD work of one of the authors (A. Hoyos), which encompassed experimental measurements of ammonia slip when a solution of ammonia, CO_2 and water is atomized in a two fluid nozzle, as well as modeling and simulation of packed column using rate-based models (work partially conducted at the Technical University of Dortmund).

1.1. Definitions

Nowadays, chemical absorption is considered as the most feasible technology to mitigate the environmental impact of CO_2 emissions in the near future. Therefore, many experimental and modeling work [1-3] has been devoted to the absorption of CO_2 with compounds such as amines, amine mixtures and ammonia, among other commercial brands.

Currently, models for gas-liquid separation process are classified as: equilibrium models (EQM), exact models (EM) and rate-based models (RBM) [4]. EQM are developed assuming equilibrium stages. That is, the chemical potential of any species in the gas and liquid streams leaving any stage are equal; the temperature of both vapor and liquid streams leaving the stage are also the same. More accurate predictions of actual behavior have been achieved by using the Murphree vapor efficiency for tray columns, and the Height Equivalent to a Theoretical Plate (HETP) for packed columns.

The EM may be developed by dividing the vapor-liquid contact zone into a large number of differential volume elements $dx.dy.dz$. Different phenomenological equations (Navier - Stokes equations for fluid flow, Fick or Maxwell-Stefan equations for species diffusion, Fourier equation for energy transfer) are then written for each element. All these equations are combined with the required mass and energy balances, rate and equilibrium equations (if reactions are taking place), and boundary conditions. In order to get an exact prediction, the number of elements in the mesh should be very large, making it difficult or impossible to solve with current workstations capacity.

The RBM, on the other hand, consider a lower number of elements than the "exact" models, while retaining the most important features in gas-liquid separations, namely, the equilibrium distribution of components between two contacted phases (at interface), the effect of convective flow, reaction rate (in reactive systems), and mixing patterns; moreover, resistance to mass diffusion and heat transfer at both sides of the interface, in the gas and liquid film, is also considered. The reactions in the model can either be developed as a separate sub-model coupled to mass and energy balances or dealt with using a proportionality constant (called enhancement factor) which takes into account the enhancing effect of reactions. The first approach is used in this chapter because it allows to account for the effects of reactions over purely physical absorption. Moreover, this approach is closer to the mechanism of reactive absorption.

The most common equipment (contactors) for absorption process include packed bed towers, tray towers and spray contactors, being the former the preferred choice for absorption of CO_2. The distinctive features of each type of contactor are as follows:

Packed tower: Due to the presence of a large number of inert pieces of a specific shape inside the tower, the liquid stream divides into numerous thin films that flow through a continuous gas phase.

Tray towers: The gas flows through small apertures located in the bottom of a tray containing a relatively thick, continuous liquid film.

Spray contactors: Liquid is dispersed into many discrete droplets within a continuous gas phase.

The main features of a packed tower for CO_2 absorption appear in Fig. **1**; diameter of the column (d_{col}) and packing height (H) are the characteristic dimensions. The tower may be filled with random or structured packing; the inlets for lean solution (ammonia, amine or other) and sour gases are located at the top and bottom, respectively, and the outlets for rich solution (solution carrying the absorbed CO_2) and the sweet gases at the bottom and top, respectively. In order to attain adequate wetting of the whole packing there is a liquid distributor at the top of the bed; additional liquid distributors should be placed if the packing height is larger than 7 meters [5]. Numbering of stages starts from top to bottom, neither condenser nor reboiler are needed for absorption towers. The other types of absorption equipment, namely tray column and spray contactors have similar configuration to packed towers but they do not have filling material and their usage is rare in absorption of CO_2.

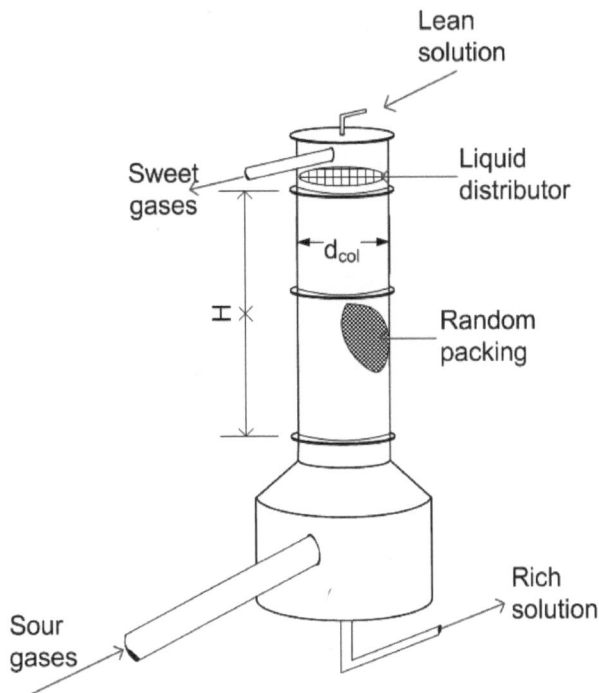

Figure 1: Basics of a packed tower.

The wide use of packed towers, at laboratory or pilot plant and plant scale, stems from their high efficiency at atmospheric pressure and low cost. There is also a large array of packing materials (plastic, ceramic or metal) and shapes (1st, 2nd or 3rd generation); besides, packing can be structured or random. The selection of the shape of the packing material determines the efficiency, diameter of the column and pressure drop, while the materials of the packing determine its cost. For example, third generation packing is more efficient than 2nd and 1st generation; 3rd generation packing will also result in lower column diameters and pressure drop. On the other hand, cost of columns with stainless steel packing is higher than that of columns with plastic or ceramic materials. In regards to structured and random packing, the former is more efficient; besides, structured packing is best suited for separations at higher pressure, whereas at atmospheric pressure random packing materials of 3rd generation are as efficient as the structured ones, with the additional advantage of lower cost. Due to the low cost and adequate performance for applications at atmospheric pressure, this chapter focuses on random packing materials.

2. SELECTION OF CONTACTOR FOR GAS ABSORPTION

As mentioned before, three types of equipments are commonly considered for absorption processes: tray, packed, and spray contactors. Packed and tray towers have to be installed vertically and operate in countercurrent flow, while the spray contactors may be installed vertically or horizontally with countercurrent or cross flow operation, respectively.

The selection of the equipment depends on the percentage of pollutants abatement required, the solubility of the gas, the reaction rates (if reactions are taking place), and the properties of the solvents, among others. As a general rule, tray columns are suitable for large installations operating with clean, non foaming, and non corrosive liquids, and medium to low liquid flow (if perforated plates are used). Packed columns, instead, are more suitable than tray columns for high liquid flow and for corrosive services, liquids tending to foam, very high liquid to gas ratios, and low-pressure drop requirements. Besides, packed columns are more versatile than tray towers and spray contactors.

On the other hand, spray contactors are preferred if the pressure drop is a critical parameter or if the process involves slurries which may plug the packing or trays;

their main disadvantage is that they are not suitable for process requiring several steps (as the absorption of CO_2) or operation close to equilibrium [6].

The selection of equipment may be based on the advantages and disadvantages described by A. L. Kohl [6] for each type of equipment. Thus, Table **1** is derived for the specific case of the absorption of CO_2 by aqueous ammonia solution; it shows the selection criteria and quantitative values assigned in the selection of the contactor type; the relative importance given to each issue is also presented in Table **1**.

Table 1: Evaluation of alternatives to select an absorption unit

Issue	Relative Importance	Spray Contactors	Packed Column	Tray Column
Availability to arrange several stages	25%	10	25	25
High capacity to manage large liquid to gas flow ratios	25%	20	20	10
High corrosion resistance	20%	15	17	5
Low reagent slip	15%	5	10	10
Low pressure drop	10%	10	10	5
Availability to manage solids	5%	5	2	1
Score		**65/100**	**84/100**	**56/100**

Since CO_2 does not absorb as easily as sulfur dioxide, the contactor chosen must be able to arrange several stages; either tray or packed columns accomplish that. Avoiding ammonia slip is also an important decision factor in the CO_2 absorption with ammonia, because ammonia is more volatile than amines; ammonia slip in spray contactors scored the lowest due to the large slip measured in a laboratory test at the Environmental Catalysis Research Group: during the tests, the ammonia slip from a solution sprayed with a two fluid nozzle was 60%.

According to the score in Table **1**, a packed tower is the best option for absorbing CO_2 with ammonia; the overall score was 84/100.

A 3rd generation random packing material was considered for the packed column in Table **1**. Although 1st and 2nd generation packing materials are also available, they are less efficient. Structured packing might be an alternative to random packing; however, random packing is easier to fabricate and to load in the column

[7]. Therefore, random packing (cheap and highly efficient near atmospheric pressure) was chosen as the best option for this gas liquid separation.

2.1. Types of Packing

Examples of random packing appear in Fig. **2**; packing generation is also indicated. A complete list of packing, including Envipac and Ralu – Flow which can also be a good choice for absorption towers due to their void fraction and their specific area, may be found in the books of Billet [8] and Maćkowiak [5].

1st generation   2nd generation   3rd generation

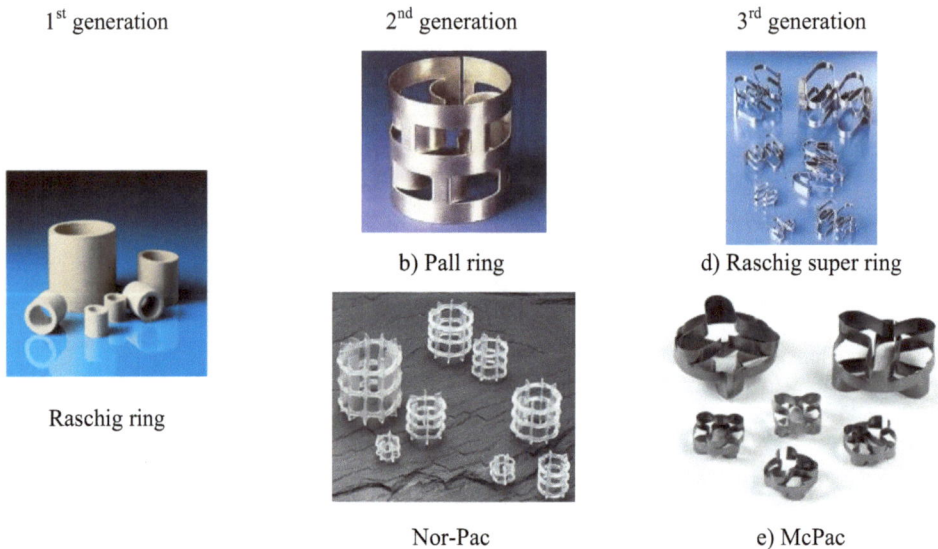

b) Pall ring   d) Raschig super ring

Raschig ring

Nor-Pac   e) McPac

Figure 2: Historic evolution of random packing material: a) Raschig ring [9], 1st generation 1930-1960; b) Pall ring [9] and c) Nor-Pac [10], 2nd generation 1960 -1980; d) Raschig super ring [9] and e) McPac [11], 3rd generation 1990.

A review of the main characteristics of some common packing materials is listed in Table **2**; it summarizes technical data of packing materials, some of them investigated by Maćkowiak [5]. Information in Table **2**, which is required for sizing packing columns (sections 4.4 and 6), includes the dimensions of each packing elements (height (H), diameter (d) and thickness (s)), the number of elements per cubic meter of packing (N), the specific area (a), the void fraction (ε) and the bulk density (G). Flooding constant ($C_{FL,0}$), constants (K1 to K4) for evaluating of resistance coefficient (ψ_{FL}) depending on the Reynolds number, the form factor (φ_P) for determining the resistance coefficient of dry packing (ψ)

and the form factor (μ) for calculating the pressure drop in irrigated packing, are also included in Table **2**.

3. COMPARISON OF RATE-BASED MODELS (NON EQUILIBRIUM MODELS) AND EQUILIBRIUM MODELS

Fluid separation process such as absorption may be modeled by rate-based models (RBM) or by equilibrium models (EQM). However, accuracy of the equilibrium models is somewhat low because streams leaving each stage are assumed to be in chemical and thermal equilibrium, which is not often the case; moreover, EQM neglects mass transfer resistances. The Murphree vapor efficiency and HETP have been implemented to improve the capabilities of EQM, but these values are unknown for a new separation unit prior to construction or without experimental tests; besides, these concepts are not readily extrapolated to the separation of more than two components.

An example problem may illustrate the differences between EQM and RBM; the problem was solved using *Aspen Plus® V7.3.*

Example: Differences between RBM and EQM.

Near 80% of CO_2 in a gaseous stream of 81,074.51 kg h^{-1}at 303.15 K and 120 kPa, is absorbed with 291,951.19 kg h^{-1} of water (lean solution) at 278.15K. The absorption is carried out in a packed column, where the vapor and liquid streams are in counter current. Packing height is 3 m, divided in 3 sections. The diameters of the sections, from top to bottom, are 2.8 m, 2.8 m and 3.0 m, Fig. **3**. All sections are filled with Ralu-Flow® No. 1 with an specific area and void fraction of 165 m^3 m^{-2} and 0.95, respectively [10]. The heat and mass transfer coefficients required in the rate-based model were calculated with the Chilton-Colburn and Billet equations, respectively, while EQM only considered the equilibrium stage, neglecting any mass or heat transfer. The EQM and the RBM approaches were tested by running the simulation under the same conditions of flow and compositions.

The results obtained by rate based-calculations and equilibrium calculations are compared in Fig. **4** in terms of temperature profiles (Fig. **4a**), mole fraction of CO_2 in vapor stream (Fig. **4b**), mole fraction of CO_2 in liquid stream (Fig. **4c**), and molar flows of vapor and liquid inside the column (Fig. **4d**).

Table 2: Characteristic parameters of some packing materials

Packing	Material	Size			$N*10^{-3}$	a	ε	G	ψ_{FL}		ψ_{FL}	
		$d*10^3$	$h*10^3$	$S*10^3$	$[1\,m^{-3}]$	m^2m^{-3}	m^3m^{-3}	Kgm^{-3}	$Re_V < 2100$		$Re_V \geq 2100$	
		[m]	[m]	[m]					K1	K2	K3	K4
MCPAC	Metal	65	30	0.4	7.7	90.6	0.982	144	5.4	-0.159	2.382	-0.052
Ralu – Flow #1	Plastic				33	165	0.95					
Ralu – Flow #2	Plastic	58	58	1.6	4.57	96	0.96	36	2.26	-0.112	1.4	-0.05
ENVIPAC	PP	60	54	2	6.8	81.6	0.957	38.8	4.792	-0.187	1.697	-0.0513
Glitsch rings CMR 304	Metal (0.5A)	16	5	--	560.8	356.8	0.955	360	6.53	-0.148	4.330	-0.0920
	Metal (1.0A)	25	8	--	160	234.7	0.971	232	5.40	-0.140	3.241	-0.0733
Nutter rings By Sulzer	Metal (No. 1.0)	--	--	0.3	67.1	168	0.978	178	11.196	-0.206	3.518	-0.051
Intalox saddles	Ceramic (1.0)	25	22	3.8	68	197	0.704	651.2	15.5	-0.255	3.800	-0.071
Intalox super saddles	Plastic	50	--	--	5.80	110	0.94	54	3.210	-0.075	3.210	-0.075
RMSR	Metal	25	--	0.4	--	235	0.96	305	7.397	-0.206	2.325	-0.051
Hackette	Plastic	42	48	2.5	12.4	135	0.93	63	10.98	-0.27	2.22	-0.065
Dtnpac	Plastic (size 1)	47	18	1.5	29.0	135.3	0.92	72	5.40	-0.159	2.382	-0.052
SR-Pac	Ceramic	65	50	5.0	4.2	105.7	0.802	437	6.6468	-0.205	1.6302	-0.022
Tellerete	Plastic	47	18	1.5	--	190	0.930	63	4.86	-0.159	2.142	-0.052

From Fig. **4a** it is noted that RBM calculates temperature profiles for vapor and liquid phases, as well as for the interface, whereas there is only one temperature (that of the liquid stream) in EQM. Indeed, EQM assumes thermal equilibrium between the liquid and vapor phase in each stage. Hence, in processes where the temperature of the phases plays an important role, such as absorption, the assumption of thermal equilibrium may be a serious disadvantage for EQM. In fact, RBM results show that thermal equilibrium is not attained in any stage. Therefore, heat transfer occurs from vapor to liquid, and the shape of the temperature curves obeys a ratio $L*Cp_L/G*Cp_G$ much larger than 1 (the value in this example is 27.8).

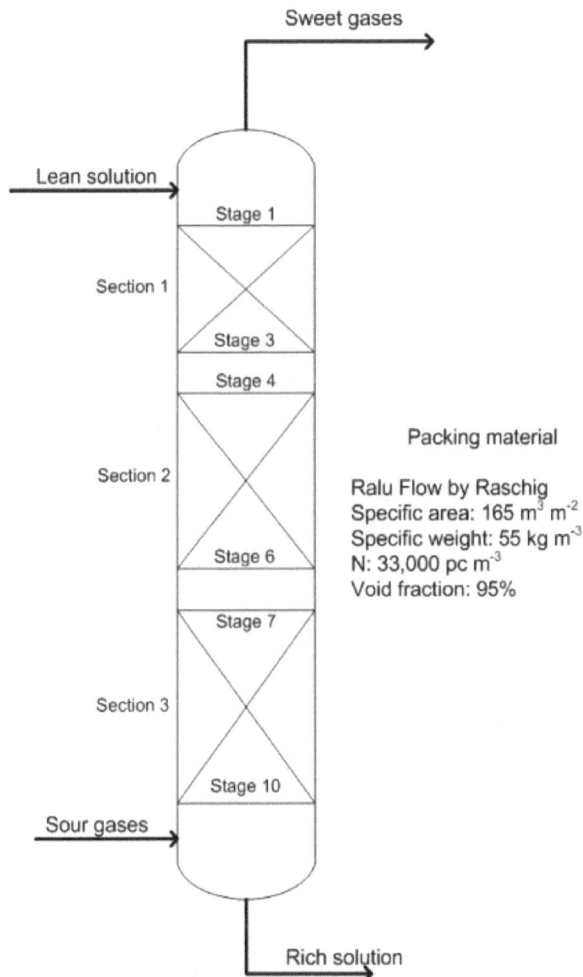

Figure 3: Sketch of a ten stages packed column (Ralu Flow packing) for absorption of CO_2 from gas mixture of CO_2 and N_2 using water.

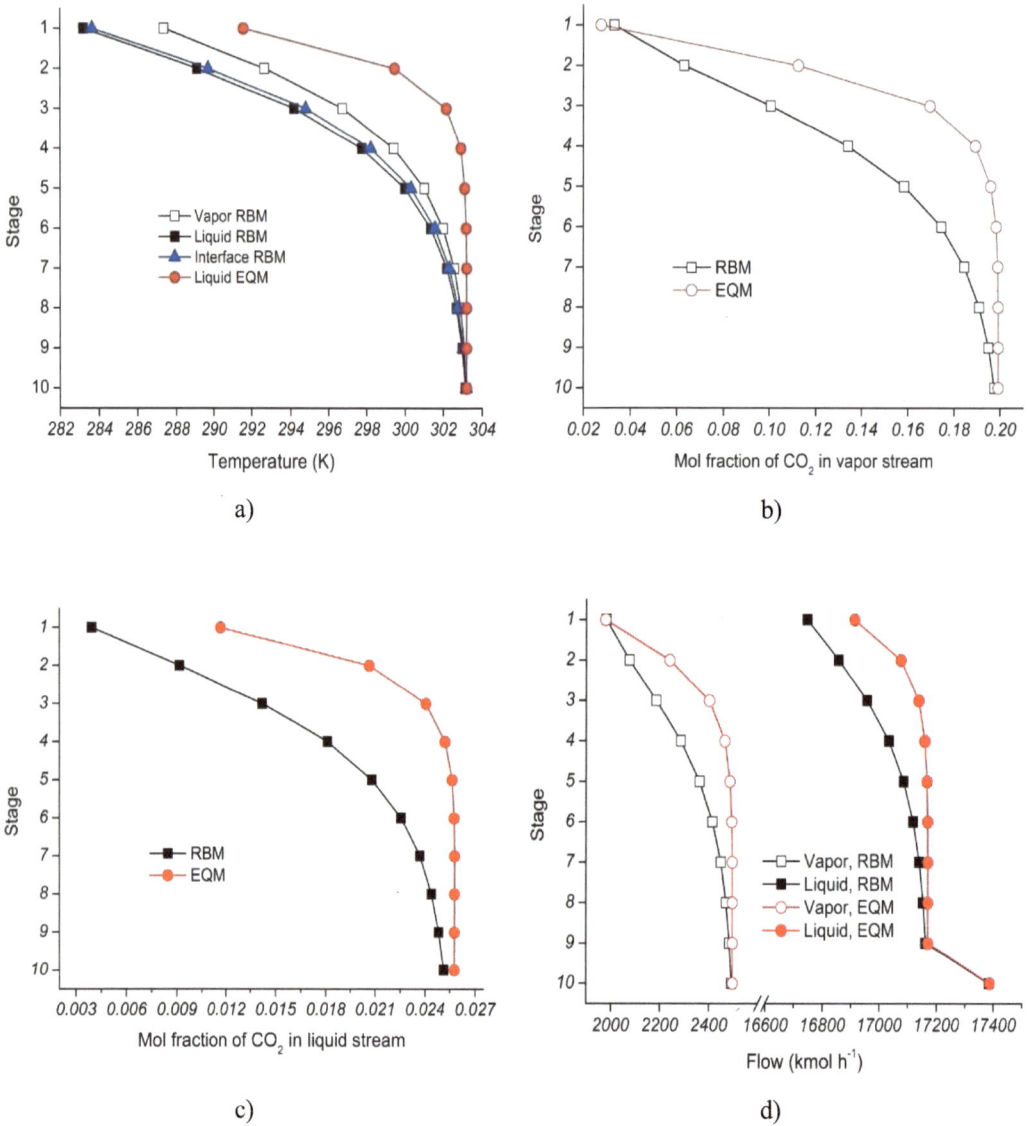

Figure 4: Comparison of RBM and EQM profiles: a) temperatures profiles of gas and liquid by RBM *vs.* temperature of liquid by EQM. b) Mole fraction of CO_2 in gas streams by RBM *vs.* EQM. c) Mol fraction of CO_2D:\For Working\E-Books\MC E-Books\E-Book\01 Asma Ahmed\355 Rosa Asma\Composed\Word Files in liquid streams by RBM *vs.* EQM. d) Molar flows of gas and liquid by RBM *vs.* EQM.

CO_2 composition in the vapor stream leaving the column (see Fig. **4b**) is very similar in equilibrium and rate-based calculations, being slightly lower in the former. However, differences in the middle section of the column are significant.

In fact, while in EQM the composition of CO_2 remains constant after the sixth stage, it changes gradually throughout the column in RBM. In addition, the trend of the two curves of gas composition (Fig. **4b**) is similar to that of the liquid temperature profiles between stages 2 and 10 (Fig. **4a**), which may be a consequence of the effect of temperature on the distribution coefficient. Thence, the profiles of CO_2 composition in liquid phase (see Fig. **4c**) also have a similar shape to liquid temperature profiles in Fig. **4**.

The flow profiles in Fig. **4d** also show considerable differences for the intermediate stages of the column. EQM predicts higher flows (liquid and vapor) than RBM and it varies in the first 6 stages and then remains constant in the last 4. Since higher liquid load results in higher pressure drops, a slightly larger column diameter would be predicted by EQM in order to attain similar pressure drops in both EQM and RBM. Other differences would be on the location of the loading zone and flooding point of the column.

Regarding the accuracy of RBM and EQM to predict experimental results, Thiele, *et al.,* [12] compared the H_2S, CO_2 and NH_3 profiles in liquid and gas phase obtained by RBM and EQM with experimental data of a packed tower. In the liquid stream, the concentration of CO_2 was best predicted by the RBM along the column: deviation between predicted and experimental results was almost negligible in RBM, whereas in EQM was as high as 10 g L^{-1} at the bottom of the column. Similarly, the experimental concentrations of H_2S and NH_3 in the liquid stream were well predicted by RBM, while the prediction by EQM showed appreciable differences at the bottom of the column.

In the gas phase, the prediction of experimental concentration of CO_2 (in ppm) by RBM was outstanding, while the deviation with EQM was around 5,000 ppm; the prediction of NH_3 and H_2S concentration was good for RBM (*i.e.,* there was no deviation in NH_3 profile and a slight deviation at the middle of the column for H_2S profile); on the other hand, deviation between experimental results and EQM for H_2S and NH_3 was significant: the concentration predicted for NH_3 was lower than the experimental result and the prediction for H_2S was larger than the experimental values (average difference along the column was around 4,000 ppm).

The striking difference between RBM and EQM may be the results of several factors: (a) Although RBM and EQM use distribution coefficients, RBM evaluates them at the interface, and EQM evaluates them at the streams leaving any stage; due to the mass transfer limitations considered in RBM, the concentration of species i at the interface may be different from the concentration in the streams leaving the stage. (b) RBM accounts for heat and mass transfer through the interface, whereas EQM does not consider either of them. (c) Turbulent and perfectly-mixed bulk liquid and gas phases are assumed in RBM, assumption that is also included in EQM; however, RBM considers that near the interface mass and heat transfer occurs by three mechanisms: drift, diffusion and convection; the drift is defined as a convective transport of mass and heat due to the movement of matter trough the interface, whereas the diffusion is a transfer mechanism due to differences in concentration and expressed by the Fick's law. Hence, in RBM the flux of a component (in a binary mixture) resulting from diffusion and drift may be expressed by equation (1).

$$N_i = -D_i \frac{dC_i}{dz} + N_{total} x_i = -k_i^{binary} * \Delta C_i + N_{tot} x_i \quad \therefore k_i^{binary} = \frac{D_i}{\delta} \ldots \tag{1}$$

Where the mass transfer coefficient (k_i) in multi-component systems could be determined from binary mass transfer coefficients as shown in equation (2).

$$k_{ij} = \frac{D_{ij}}{\delta} \ldots \tag{2}$$

Alternatively, the mass transfer through the interface may be described more accurately by Maxwell-Stefan equations (3), which consider the chemical potential gradient as the driving force which causes the motion of species i, which is compensated by j species.

$$\frac{d\mu_i}{dx} = \sum_{j=1}^{n} x_j \zeta_{ij} (v_j - v_i) \ldots \tag{3}$$

The EQM, on the other hand, does not consider explicitly mass transfer, *i.e.*, equations (1) – (3), but relies only on the distribution coefficients (K_i) to describe the transference of species i between two phases in contact.

4. DEVELOPING A RBM

Although the equilibrium model is quite good for describing the distillation of ideal binary mixtures with close boiling points, the RBM better describes the absorption process, because of its rigorous treatment of the reactive system and electrolyte mixtures. As mentioned before, RBM considers absorption processes as rate-controlled processes where the mass and heat transfer limitations play an important role in the overall process of gas – liquid separation. Thence, this section is aimed at describing the equations for a RBM.

Fig. **5** is a representation of a gas-liquid separation process in a column [13]. Fig. **5a** shows the mass and energy transfer in a non-equilibrium stage for a RBM. In this representation, the composition of the streams leaving the stage are determined by computing the mass and energy transfer across the interface in any stage; it is worth noting that this is one of the key differences from the EQM where the assumption that composition of vapor stream leaving the stage j $\left(y_{ij}\right)$ is in equilibrium with the composition of the liquid stream leaving stage j $\left(x_{ij}\right)$ is the basis for the calculations. According to Fig. **5a**, each segment could have a feed stream$\left(F_j z_{i,j}\right)$, gas $\left(U_j y_{i,j}\right)$ and liquid $\left(W_j x_{i,j}\right)$ side streams, and an interface between liquid and vapor films. E_j and N_{ij} represent the heat transfer and mass transfer of species i through the interface in stage j, respectively.

Fig. **5b** depicts a multi-stage separation unit; the stages are numbered from top to bottom and side streams are included in each stage. This is a general representation which may vary depending on the process. For example, unlike distillation, absorption processes rarely include side streams (U and W) and, consequently, only the top and bottom stages include inlet streams for liquid and vapor (gas), respectively. On the other hand, discretization of the liquid film in each stage is required in absorption processes due to the large deviation from the condition of constant molar flow and the possible presence of mass transfer limitations and reactions; such discretization is not necessary in RBM for distillation operations.

The mass and heat balances for a RBM are formulated in separated phases, assuming that the bulk phases are well mixed. In addition, mass and heat transfer limitations are included in the balance equations for the gas and liquid films at both sides of the interface; mass transfer correlations are used to evaluate mass transfer limitations, which depend on the internal streams velocity, dimension of the column and type of packing.

As a consequence, from a modeling point of view it is more convenient to treat the feed, top and bottom stages separately, dividing each stage in gas bulk, gas film, interface, liquid film and liquid bulk. This will be discussed in the next section.

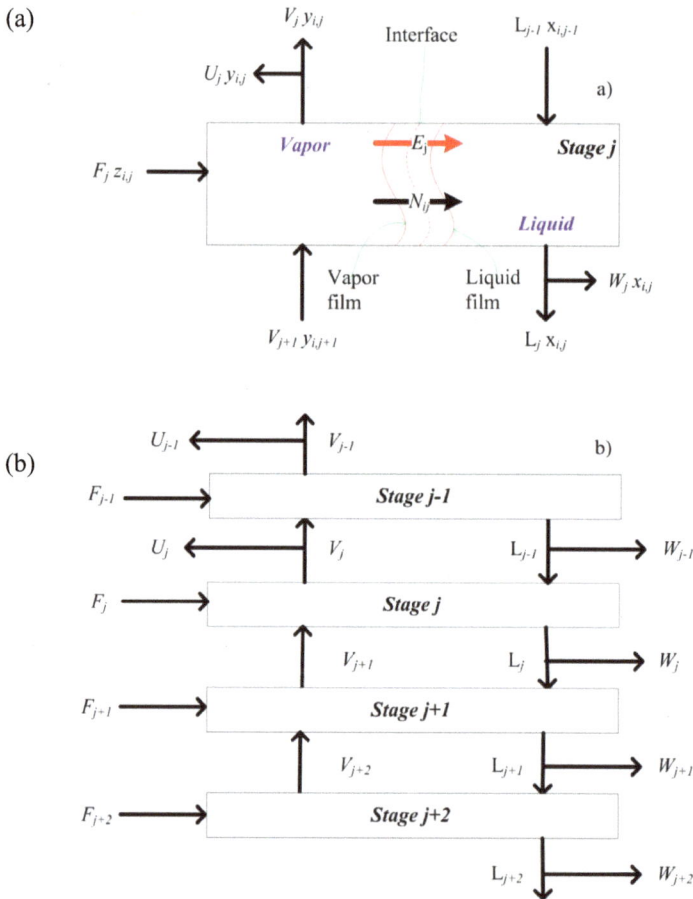

Figure 5: (a) Representation of non-equilibrium stage (b) Multi-stage separation column.[1]

---

[1] Adapted from: Perry's Chemical Engineering Handbook 8th edition, Figs. **13-54** and Fig. **13-36**

4.1. Steady-State Mass Balance in the Bulk

The assumption of well-mixed bulk phases with homogeneous concentration is best achieved if a large enough number of segments is used. Mass balances for the vapor and liquid phases in each stage are given by equations (4) and (5), respectively.

$$\left(1+r_j^V\right)V_j y_{i,j} - V_{j+1}y_{i,j+1} - f_{i,j}^V + N_{i,j}^V + R_{i,j}^V = 0 \tag{4}$$
$$\ldots$$

$$\left(1+r_j^L\right)L_j x_{i,j} - L_{j-1}x_{i,j-1} - f_{i,j}^L - N_{i,j}^L + R_{i,j}^L = 0 \tag{5}$$
$$\ldots$$

Both equations include internal flows (V_j, L_j), side streams (included in r_j), feed to the stage ($f_{i,j}$), mass flow though the interface ($N_{i,j}$), and one term for reactions ($R_{i,j}$) if they are present. In absorption process, the side stream ratio $\left(r_j\right)$ and feed streams $\left(f_{i,j}\right)$ in intermediate stages are zero.

4.2. Steady-State Energy Balance in the Bulk

Similarly to mass transfer, heat balance equations in terms of the enthalpy of streams entering and leaving each phase have to be formulated separately for each phase, equations (6) and (7). Heat transferred through the interface, heat transferred to each phase, and heat of reaction (if reactions are taking place in any of the phases) are also included in the heat balances.

$$\left(1+r_j^V\right)V_j H_j^V - V_{j+1}H_{j+1}^V - F_j^V H_j^{VF} + E_j^V + Q_j^V + \Delta H_j^V = 0 \ldots \tag{6}$$

$$\left(1+r_j^L\right)L_j H_j^L - L_{j-1}H_{j-1}^L - F_j^L H_j^{LF} - E_j^L + Q_j^L + \Delta H_j^L = 0 \tag{7}$$
$$\ldots$$

When equations (6) and (7) are applied for absorption process, the side stream ratio and feed to intermediate stages are usually zero.

4.2. Mass and Energy Balances in the Interface

4.2.1. Mass Balance at the Interface

Continuity of molar flow across the interface, equation (8), is the building block of the mass balance equations. Molar flows are evaluated from molar fluxes and interfacial area, equation (9). Molar fluxes from the bulk (vapor or liquid) to the interface are a function of the concentration gradient between the bulk and the interface, mass transfer coefficients, and concentration of the bulk (vapor or liquid) phase, equations (10) and (11). Finally, vapor and liquid concentrations at the interface are calculated from the phase equilibrium expressed in terms of distribution coefficients, equation (12). Even though *the form* of the equation for phase equilibrium is exactly the same used for EQM, in RBM the concentrations do not refer to the streams leaving the stage but to the interface, which does not necessarily imply that streams leaving the stage will be in phase equilibrium.

$$N_{i,j}^V = N_{i,j}^L \dots \tag{8}$$

$$N_{i,j}^p = \int N_{i,j}^p da_j \dots \tag{9}$$

$$N_i^V = c_t^V k_i^V \left(y_i^V - y_i^I \right) + y_i^V N_t^V \dots \tag{10}$$

$$N_i^L = c_t^L k_i^L \left(x_i^I - x_i^L \right) + x_i^L N_t^L \dots \tag{11}$$

$$y_i^I = K_i x_i^I \dots \tag{12}$$

4.2.2. Energy Balance in the Interface

The main contribution to the energy balance is heat transfer. These equations are set up in terms of the energy balance at the interface, equation (13), the energy flow from the bulk to the interface, equation (14), and the heat fluxes from the vapor and liquid phases, equations (15) and (16), respectively. One of the characteristics of RBM is that the temperature of the interface approximates that of one of the two phases; for instance, in the previous example the temperature of the interface was close to that of the liquid stream. The approach of the interface temperature to liquid or gas stream will depend on heat transfer coefficients (h^V and h^L) in equations (17) and (18), which are different from zero for RBM.

$$E_j^V = E_j^L \dots \tag{13}$$

$$E_j^p = \int E_j^p da_j \dots \tag{14}$$

$$E^V = q^V + \sum_{i=1}^{c} N_i^V H_i^V = E^L \ldots \tag{15}$$

$$E^L = q^L + \sum_{i=1}^{c} N_i^L \overline{H}_i^L = E^V \ldots \tag{16}$$

$$q^V = h^V \left(T^V - T^I \right) \ldots \tag{17}$$

$$q^L = h^L \left(T^I - T^L \right) \ldots \tag{18}$$

Mass and heat transfer coefficients in the liquid and vapor streams are determined by the flows of the streams and its ratio, the dimensions of the column, and packing characteristics; all these parameters are included in hydrodynamic and mass transfer correlations.

4.3. Hydrodynamics of Randomly Packed Columns

The hydraulic behavior of a two-phase countercurrent gas-liquid or vapor-liquid packed column can be explained by the relationship between the pressure drop Δp per unit length and the velocity u_V or capacity factor $F_V = u_V * \sqrt{\rho_V}$ of gas or vapor flowing through the column. The relationship between capacity factor, pressure drop and liquid load also determines the liquid hold-up and hence the separation efficiency. Fig. **6** provides a qualitative illustration of the relationship of pressure drop $\left(\Delta p / H \right)$ and liquid hold-up (h_L) with the vapor capacity factor (F_V) for different values of liquid load u_L.

In Fig. **6**, segment $\overline{AA}$, which is located at about 65% of the flooding capacity factor $F_{V,Fl}$, is called the *loading line* [5]; segment $\overline{BB}$ corresponds to the *upper loading line*, and segment $\overline{CC}$ represents the *flooding line*, which is reached when the gas capacity factor is equal to the flooding capacity factor. Below the loading line, the liquid hold-up is independent of the gas capacity factor, and hence, independent of gas velocity, as shown in the plot of h_L vs. F_V. In the range where liquid hold-up is independent of F_V, flooding will not occur in spite of variations in gas velocity, but the efficiency in this operation zone is small.

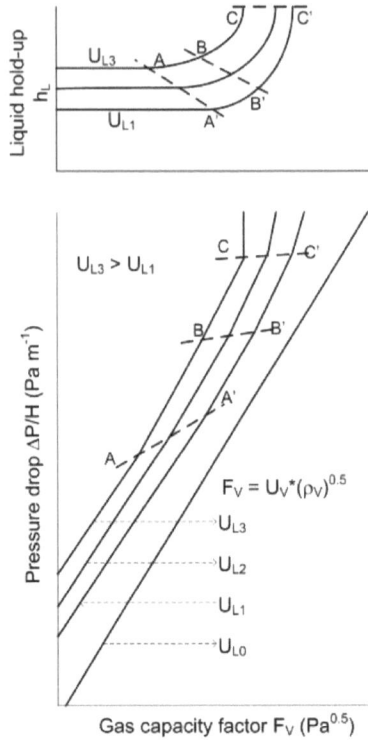

Figure 6: Liquid hold-up (up) and pressure drop (bottom) in packed column as function of the gas capacity factor.[2]

As F_V increases (starting from the loading line $\overline{AA}$), the liquid hold-up reaches the upper loading line ($\overline{BB}$). Operation at the upper loading line ensures reaching the maximun efficiency (for a dicussion of efficiency as function of gas capacity factor see Maćkowiak [14]). Further increments of F_V will result in pinching the flooding line ($\overline{CC}$), where a sligh increment of F_V will result in column flooding.

The upper loading and the flooding lines determine the range of safe operation of the column; the ideal working condition is around the upper loading line. Other important factor is the gas capacity factor; it should be larger than 1.0 for a randomly packed column of reactive absorption at atmospheric pressure.

Models for the prediction of the hydrodynamic behaviour have been reported by Billet and Schultes [15-17] and Maćkowiak [5, 14]. The equations for calculating

---

[2] Adapted from: Fluid Dynamics of Packed Columns: Principles of the Fluid Dynamics Design of Column for Gas/Liquid and Liquid/Liquid Systems, chapter two, pg. 26.

the pressure drop and the liquid holdup according to the model proposed by Billet and Schultes are summarized in Table **3**.

Table 3: Hydrodynamic correlations from Billet and Schultes model [17]

Parameter	Application Zone	Correlation		
Pressure drop	Untrickled packing	$\dfrac{\Delta p_0}{H} = \psi_0 \dfrac{a}{\varepsilon^3}\dfrac{F_V^2}{2}\dfrac{1}{K}$ (19)	$\dfrac{1}{K} = 1 + \dfrac{2}{3}\dfrac{1}{1-\varepsilon}\dfrac{d_p}{d_s}$ (20)	$\psi_0 = C_{p,0}\left(\dfrac{64}{Re_V} + \dfrac{1.8}{Re_V^{0.08}}\right)$ (21)
		$Re_V = \dfrac{u_V d_p}{(1-\varepsilon)v_V}K$ (22)	$d_p = 6\dfrac{(1-\varepsilon)}{a}$ (23)	
Pressure drop	Irrigated packing below the loading point	$\dfrac{\Delta p}{H} = \psi_L \dfrac{a}{(\varepsilon - h_L)^3}\dfrac{F_V^2}{2}\dfrac{1}{K}$ (24)	$\psi_L = C_{p,0}\left(\dfrac{64}{Re_V} + \dfrac{1.8}{Re_V^{0.08}}\right)\left(\dfrac{\varepsilon - h_L}{\varepsilon}\right)^{1.5}\left(\dfrac{h_L}{h_{L,s}}\right)^{0.3}\exp\left(C_1\sqrt{Fr_L}\right)$ with $C_1 = \dfrac{13300}{a^{3/2}}$ (25)	
		$Fr_L = \dfrac{u_L^2 a}{g}$ (26)		
Phases velocities	Loading point	$u_{V,S} = \sqrt{\dfrac{g}{\psi_S}}\left[\dfrac{\varepsilon}{a^{1/6}} - a^{1/2}\left(12\dfrac{1}{g}\dfrac{\eta_L}{\rho_L}u_{L,S}\right)^{1/3}\right]\times \left(12\dfrac{1}{g}\dfrac{\eta_L}{\rho_L}u_{L,S}\right)^{1/6}\times\sqrt{\dfrac{\rho_L}{\rho_V}}$ (27)	$\psi_S = \dfrac{g}{C_S^2}\left[\dfrac{L}{V}\sqrt{\dfrac{\rho_V}{\rho_L}}\left(\dfrac{\eta_L}{\eta_V}\right)^{0.4}\right]^{-2n_S}$ for $\dfrac{L}{V}\sqrt{\dfrac{\rho_V}{\rho_L}} \le 0.4 : n_S = -0.326; C_S \to \text{table 4}$ for $\dfrac{L}{V}\sqrt{\dfrac{\rho_V}{\rho_L}} \ge 0.4 : n_S = -0.723; C_S = 0.695 C_S\left(\dfrac{\eta_L}{\eta_V}\right)^{0.1588}$ (28)	
		$u_{L,S} = \dfrac{\rho_V}{\rho_L}\dfrac{L}{V}u_{V,S}$ (29)		
Phases velocities	Flooding point	$u_{V,Fl} = \sqrt{2}\sqrt{\dfrac{g}{\psi_{Fl}}}\dfrac{(\varepsilon - h_{L,Fl})^{3/2}}{\varepsilon^{1/2}}*\sqrt{\dfrac{h_{L,Fl}}{a}}\sqrt{\dfrac{\rho_L}{\rho_V}}$ (30)	$\psi_{Fl} = \dfrac{g}{C_{Fl}^2}\left[\dfrac{L}{V}\sqrt{\dfrac{\rho_V}{\rho_L}}\left(\dfrac{\eta_L}{\eta_V}\right)^{0.2}\right]^{-2n_{Fl}}$ for $\dfrac{L}{V}\sqrt{\dfrac{\rho_V}{\rho_L}} \le 0.4 : n_{Fl} = -0.194; C_{Fl} \to \text{table 4}$ for $\dfrac{L}{V}\sqrt{\dfrac{\rho_V}{\rho_L}} \ge 0.4 : n_{Fl} = -0.708; C_{Fl} = 0.6244 C_{Fl}\left(\dfrac{\eta_L}{\eta_V}\right)^{0.1028}$ (31)	
		$u_{L,Fl} = \dfrac{\rho_V}{\rho_L}\dfrac{L}{V}u_{V,Fl}$ (32)		

In Table **3**, expressions for the pressure drop in untrickled and irrigated packings are presented. In the former, pressure drop, equation (19), depends on the geometric surface (a) and void volume of the packing (ε), the gas capacity factor (F_V), the wall factor (K), which is related to the increased void fraction at the column wall, and the resistance coefficient (ψ_0). The wall factor, resistance coefficient, Reynolds number, and particle diameter can be calculated with equations (20) to (23). When the untrickled packing is irrigated, on the other hand, the free cross section for the gas flow is reduced by the column hold-up. Moreover, the surface parameters change as a result of the coating with the liquid film. Therefore, the pressure drop is calculated by equation (24), where the resistance coefficient depends on the vapor Reynolds number, void fraction, liquid hold-up, and Froude number. The resistance coefficient and Froude number can be calculated by equations (25) and (26), respectively.

As the operation regime of the column is bounded by the loading and flooding points, Table **3** also shows expressions for the vapor and liquid velocities at those conditions. At the loading point, the vapor velocity, calculated by equation (27), depends on the resistance coefficient (ψ_S), which in turn depends on the mass flow ratio (L/V); the (L/V) ratio is also related to the phase densities. The n_S exponent in equation (28) is a function of the phase inversion parameter $\left(L/V \sqrt{\rho_V / \rho_L} \right)$: if the phase inversion parameter is greater than 0.4, phase inversion occurs and the liquid flows downwards as a continuous phase, while the vapor phase rises up through the liquid layer in the form of bubbles. In such case, the packing specific constant (C_S) read from Table **4** has to be corrected through the equation for (C_S).

Table 4: Constants for dumped packing

Packing		Size (mm)	C_S	C_{Fl}	C_h	$C_{p,0}$	C_L	C_V
Raschig Super-Ring	Metal	0.3	3.560	2.340	0.750	0.760	1.500	0.450
		0.5	3.350	2.200	0.620	0.780	1.450	0.430
		1	3.491	2.200	0.750	0.500	1.290	0.440
		2	3.326	2.096	0.720	0.464	1.323	0.400
		3	3.260	2.100	0.620	0.430	0.850	0.300
	Plastic	2	3.326	2.096	0.720	0.377	1.250	0.337
Ralu Flow	Plastic	#1	3.612	2.401	0.640	0.485	1.486	0.360
		#2	3.412	2.174	0.640	0.350	1.270	0.320
Envi Pac	Plastic	80	2.846	1.522	0.641	0.358	1.603	0.257
		60	2.987	1.864	0.794	0.338	1.522	0.296
		32	2.944	2.012	1.039	0.549	1.517	0.459

Taken from Billet and Schultes [17].

When the gas velocity is increased above the loading point the flow of liquid downwards the column is reduced, which results in an increase in the liquid holdup. Eventually, if the gas velocity is very high, the entire liquid may be entrained by the gas phase and transported to the top of the column, condition corresponding to the flooding point. At the flooding point, the velocities of both phases are calculated by equations (30) to (32). The liquid hold-up to be used in equation (30) corresponds to the hold-up calculated by equation (55), and the packing specific constant (C_S) is shown in Table **4** for different packing.

Correlations developed by Maćkowiak for calculating the vapor velocity, pressure drop, and liquid holdup at flooding are based on the "suspended bed of droplets" (SBD) model. This model is accurate (*i.e.*, accuracy $< \pm 12\%$) for phases flow ratios in the range $10^{-4} < \lambda_0 < 1$.

In order to determine the diameter of a packed column, the model first calculates the phase flow ratio by equation (33). The liquid hold-up is then calculated by equation (34), applicable for Reynolds number larger than 2: $(Re_L \geq 2)$. Subsequently, the vapor velocity at flooding, which depends on packing characteristics, phase densities and liquid hold-up at flooding, can be calculated by equation (35). Finally, the gas capacity factors, at the flooding and loading point, can be calculated by equations (36) and (37), respectively. In all equations Reynolds number for vapor and liquid phases, which depends on the phase velocity, phase properties and packing characteristics, can be calculated by equations (38) and (39).

Similarly to the description proposed by Billet and Schultes, Maćkowiak's model calculates the pressure drop for dry packing, and irrigated packing below the loading point, up to the flooding point and at the flooding point, equations (40) to (43). Additionally, liquid holdup can be calculated by equation (44) if the column operates below the loading point and by equation (45) if the column is operating above the loading point and below the flooding point. Auxiliary calculations for droplet size, wall coefficient, hydraulic diameter of the packing element, and the resistance coefficient, are also included in the model, equations (46) to (49).

4.4. Mass Transfer Correlations

A thorough review of mass transfer correlations for randomly packed columns is presented by Wang [18]. In that work the evolution of mass transfer correlations, starting from the contributions of Sherwood, Van Krevlen and Hoftijzer, Morris and Jackson, and Shulman is presented. The paper also shows the most recent correlations developed by Onda, Cornel, Fair, Zech and Mersman, Billet and Schultes, Wagner, and Piché. In this chapter we use the correlations developed by Billet and Schultes, Onda, and Mersman, which are the most commonly used expressions. Although the correlations developed by Piche, which are based on Artificial Neural Networks, are appealing to predict the mass transfer coefficient of randomly packed columns, they were excluded from the scope of this chapter.

Table 5: Hydrodynamic correlations from Maćkowiak model

Parameters	Zone of application	Correlations				
Liquid holdup	Flooding point	$\lambda_0 = \dfrac{u_L}{u_{V,Fl}}$ (33)		$h^0_{L,Fl} = \dfrac{\sqrt{1.44 * \lambda_0^2 + 0.8 * \lambda_0 * (1-\lambda_0)} - 1.2 * \lambda_0}{0.4 * (1-\lambda_0)}$		(34)
		$u_{V,Fl} = 0.8 * \cos(\pi/4) * \dfrac{\varepsilon^{\left(\frac{6}{5}\right)}}{\psi^{\frac{1}{6}}} * \sqrt{\dfrac{d_T * (\rho_L - \rho_V) * g}{\rho_V}} \left(\dfrac{d_h}{d_T}\right)^{\frac{1}{4}} * (1 - h^0_{L,Fl})^{\frac{7}{2}}$ (35)			$F_{V,Fl} = u_{V,Fl} * \sqrt{\rho_G}$	(36)
		$F_V = u_V * \sqrt{\rho_V}$ (37)	$Re_V = \dfrac{u_V * d_p}{(1-\varepsilon) * \eta_V} * K = \dfrac{u_V * d_p}{(1-\varepsilon) * (\eta_V)} * $ (38)		$Re_L = \dfrac{u_L * \rho_L}{a * \eta_L} = \dfrac{u_L * \rho_L}{a * (\eta_L)}$	(39)
Pressure drop	Dry and pre-loading	$\dfrac{\Delta P}{H} = \psi_0 * \dfrac{1-\varepsilon}{\varepsilon^3} * \dfrac{F_V^2}{d_p * K}$ (40)		$\dfrac{\Delta P}{H} = \psi_0 * (1 - \phi_P) * \dfrac{1-\varepsilon}{\varepsilon^3} * \dfrac{F_V^2}{d_p * K} * \left(1 - \dfrac{0.4}{\varepsilon_0} * a_0^{\frac{1}{3}} * u_L^{\frac{2}{3}}\right)^{-5}$		(41)
	Loading and flooding	$\dfrac{\Delta P}{H} = \psi_0 * (1 - \phi_P) * \dfrac{1-\varepsilon}{\varepsilon^3} * \dfrac{F_V^2}{d_p * K} * \left(1 - \dfrac{C_{BS}}{\varepsilon_0} * a_0^{\frac{1}{3}} * u_L^{\frac{2}{3}}\right)^{-5}$ (42)		$\dfrac{\Delta P}{H} = \psi_{FL} * \dfrac{1-\varepsilon}{\varepsilon^3} * \dfrac{F_V^2}{d_p * K} * \left(1 - \dfrac{0.407}{\varepsilon_0} * \lambda_0^{-0.16} * a_0^{\frac{1}{3}}\right)$		(43)
Liquid hold-up	Loading and preloading	$h_L = 0.57 * F_{r,L}^{\frac{1}{3}}$ (44)		$h_{L,S} = Z * h_L$		(45)
Auxiliary equations		$d_P = 6 * \dfrac{1-\varepsilon}{a}$ (46)	$\dfrac{1}{K} = 1 + \dfrac{2}{3} * \dfrac{\varepsilon}{1-\varepsilon} * \dfrac{d_p}{d_{col}}$ (47)	$d_h = \dfrac{4 * \varepsilon}{a} * K$ (48)	$\psi_0 = \dfrac{725.6}{Re_V} + 3.203$	(49)

4.4.1. Mass Transfer Correlations by Billet and Schultes

These correlations, which are in good agreement with one of the largest experimental data collection available of mass transfer on packed columns, are based on two major assumptions. Firstly, the empty space of dumped or arranged packings can be described by vertical flow channels: in each channel, liquid trickles downwards as evenly distributed drops while the gas flows upwards in countercurrent flow. Secondly, the deviation of the real flow behavior of the phases from the vertical flow channels approximation can be expressed by a packing-specific shape constant [17]. According to Billet and Schultes the volumetric mass transfer coefficients are given by equations (50) and (51).

$$k_{L\_Billet} * a_{ph} = C_{L\_Billet} * 12^{\frac{1}{6}} * \left(\frac{u_L}{h_L}\right)^{\frac{1}{2}} * \left(\frac{D_{L,i} * 10^{-4}}{d_h}\right)^{\frac{1}{2}} * a \left(\frac{a_{ph}}{a}\right) \quad \ldots \tag{50}$$

$$k_{G\_Billet} * a_{ph} = C_{V\_Billet} * \frac{1}{(\varepsilon - h_L)^{\frac{1}{2}}} * \left(\frac{a}{d_h}\right)^{\frac{1}{2}} * (D_{V,i} * 10^{-4}) * \left(\frac{u_V * \rho_V * 10^3}{a * \eta_V^*}\right)^{\frac{3}{4}} * \quad \ldots \tag{51}$$

$$\left(\frac{\eta_V^*}{D_{V,i} * 10^{-4} * \rho_V * 10^3}\right)^{\frac{1}{3}} * a_{ph}$$

These mass transfer coefficients depend on: vapor and liquid density (ρ_V, ρ_L), vapor and liquid dynamic viscosity (η_V, η_L), liquid surface tension (σ_L), vapor and liquid flow velocities (u_V, u_L), the specific surface area of the packing (a) and the void faction of the packing (ε), the Billet' specific constant for packing $(C_{V\_Billet}, C_{L\_Billet})$, which are listed in Table **4**, and the liquid hold-up (h_L). If the velocity of vapor phase is lower than or equal to the velocity at the loading point $(u_{V,s})$, that is, the velocity at any point in the segment $\overline{AA}$ in Fig. **6**, the liquid hold up can be calculated by equation (52).

$$h_L = \left(12 \frac{1}{g} \frac{\eta_L}{\rho_L} u_L a^2\right) \quad for \quad u_V \leq u_{V,s} \ldots \tag{52}$$

The corresponding specific interfacial area (a_{ph}) at this vapor velocity can be determined by equation (53).

$$a_{ph} = a * 1.5 * (a * d_h)^{-0.5} * \left(\frac{u_L * d_h * \rho_L}{\eta_L * 10^{-3}}\right)^{-0.2} \quad \ldots \tag{53}$$

$$* \left(\frac{u_L^2 * \rho_L * d_h}{\sigma_L}\right)^{0.75} * \left(\frac{u_L^2}{g * d_h}\right)^{-0.45}$$

Above the loading point, the liquid hold-up and the interfacial area increase and the effective flow velocity of the falling film decreases. In such cases, the liquid hold-up (h_L) and the specific interface area (a_{ph}/a) can be calculated by equations (54) and (56), respectively.

$$h_L = h_{L,s} + \left(h_{L,Fl} - h_{L,s}\right)\left(\frac{u_V}{u_{V,Fl}}\right)^{13} \dots \tag{54}$$

$$h_{L,Fl}^3\left(3h_{L,Fl} - \varepsilon\right) = \frac{6}{g}a^2\varepsilon\frac{\eta_L}{\rho_L}\frac{L}{V}\frac{\rho_V}{\rho_L}u_{V,Fl} \quad \text{with} \quad \frac{\varepsilon}{3} \leq h_{L,Fl} \leq \varepsilon \tag{55}$$

$$\frac{a_{ph}}{a} = \frac{a_{ph,s}}{a} + \left(\frac{a_{ph,Fl}}{a} - \frac{a_{ph,s}}{a}\right)\left(\frac{u_V}{u_{V,Fl}}\right)^{13} \dots \tag{56}$$

$$\frac{a_{ph,Fl}}{a} = 10.5\left(\frac{\sigma_L}{\sigma_W}\right)^{0.56}\left(ad_h\right)^{-0.5}\left(\frac{u_L d_h}{v_L}\right)^{-0.2}\left(\frac{u_L^2 \rho_L d_h}{\sigma_L}\right)^{0.75}\left(\frac{u_L^2}{gd_h}\right)^{-0.45} \dots \tag{57}$$

$$\bar{u}_L = \left(\frac{g\rho_V^2 u_V^2}{12\eta_L a^2 \rho_L}\right)^{1/3}\left(\frac{L}{V}\right)^{2/3}\left[1 - \left(\frac{u_V - u_{V,s}}{u_{V,Fl} - u_{V,s}}\right)^2\right] \tag{58}$$

for $\quad u_{V,s} \leq u_V \leq u_{V,Fl}$

The equations above are accurate for neutral or positive systems. If the surface tension of the liquid along the column decreases from top to bottom the process is a negative system; then, the Marangoni effect must be considered.

In actual packed columns the assumption of the presence of permanent channels whose surface is completely wetted by liquid may not be valid. As a result, the theoretical column hold-up deviates from the actual column hold-up. Comparison with experimental data indicates that deviation between the theoretical and the real column hold-up can be expressed by means of a hydraulic surface area of packing (a_h). Mathematical expressions for determining the real hold-up below the loading point are given by equations (59) to (62).

$$h_L = \left(12\frac{1}{g}\frac{\eta_L}{\rho_L}u_L a^2\right)^{1/3}\left(\frac{a_h}{a}\right)^{2/3} \dots \tag{59}$$

$$\mathrm{Re}_L = \frac{u_L \rho_L}{a \eta_L} < 5: \quad \frac{a_h}{a} = C_h \left(\frac{u_L \rho_L}{a \eta_L} \right)^{0.15} \left(\frac{u_L^2 a}{g} \right)^{0.1} \cdots \tag{60}$$

$$\mathrm{Re}_L = \frac{u_L \rho_L}{a \eta_L} \geq 5: \quad \frac{a_h}{a} = C_h 0.85 \left(\frac{u_L \rho_L}{a \eta_L} \right)^{0.25} \left(\frac{u_L^2 a}{g} \right)^{0.1} \cdots \tag{61}$$

$$h_{L,Fl} = 2.2 h_L \left(\frac{\eta_L \rho_W}{\eta_W \rho_L} \right)^{1/3} \cdots \tag{62}$$

Above the loading point, the prediction of the actual hold-up is made by equation (54), with the column hold-up $(h_{L,s})$ calculated by equation (59) and the $h_{L,Fl}$ calculated by equation (62).

4.4.2. Mass Transfer Correlations by Onda

The empirical model proposed by Onda [19] uses a different approach to estimate the interfacial area and, consequently, to calculate the vapor- and liquid-side mass transfer coefficients. Interfacial area, in this model, is replaced by the wetted area, equation (63), where the effective wetted area (a_w/a) depends on the surface tension of the liquid and the Reynolds number. This equation has been found to yield errors below 22% for columns packed with Raschig rings, Berl saddles, and spheres and rods made of ceramic, glass and polyvinylchloride. Onda *et al.*, validated their mass transfer correlations in the vaporization of water and absorption of a gas in organic solvents.

$$a_w = a * \left(1 - \exp \left[\begin{array}{l} -1.45 * \left(\dfrac{\sigma_C}{\sigma_L} \right)^{0.75} * \left(\dfrac{\rho_L * u_L}{a * \mu_L^*} \right)^{0.1} \\[2mm] * \left(\dfrac{u_L^2 * a}{g} \right)^{-0.05} * \left(\dfrac{\rho_L * u_L^2}{\sigma_L * a} \right)^{0.2} \end{array} \right] \right) \cdots \tag{63}$$

$$a_w = a * \left(1 - \exp \left[-1.45 * \left(\frac{\sigma_C}{\sigma_L} \right)^{0.75} * (\mathrm{Re}_L)^{0.1} * (Fr)^{-0.05} * (We)^{0.2} \right] \right)$$

These equations are limited to liquid Reynolds number between 0.04 and 500, Webber number between $1.2*10^{-8}$ and 0.272, Froude number between $5*10^{-9}$ and $1.8*10^{-2}$, and critical surface tension of packing of 0.061 or $0.03 < \sigma_C/\sigma_L < 2$.

The liquid-side mass transfer from Onda's correlation is given by equation (64), where the Reynolds number is calculated using the wetted area (a_w) instead of the total area of packing (a). The resulting equation for the liquid-side mass transfer is dependent on the liquid properties (density, viscosity and diffusivity) and the packing characteristics, such as total area (a), diameter of packing element (d_p) and the wetted surface area (a_w). The density in this equation is expressed in kg m⁻³.

$$k_{L,ondacorr} = 0.0051*\left(\frac{u_L*\rho_L}{a_w*\mu_L^*}\right)^{2/3}*\left(\frac{\mu_L^*}{\rho_L*D_L}\right)^{-0.5}*(a*d_p)^{2/5}\left(\frac{g*\mu_L}{\rho_L}\right)^{\frac{1}{3}}... \tag{64}$$

The gas-side mass transfer correlation $(k_{G\_ondacorr})$, equation (65), is also dependent on Reynolds number evaluated at the gas-phase conditions.

$$k_{G\_ondacorr} = 5.23*\left(\frac{a*D_G}{RT}\right)*\left(\frac{u_G*\rho_G}{a*\mu_G^*}\right)^{0.7}*\left(\frac{\mu_G^*}{\rho_G*D_G}\right)^{\frac{1}{3}}*(a*d_p)^{-2}... \tag{65}$$

4.4.3. Mass Transfer Correlations by Zech/Mersmann

Zech and Mersmann developed a method to calculate the liquid-phase mass-transfer coefficient for packed columns using the penetration theory [18]. The mass transfer coefficients are based on the calculation of the interfacial area, given by equation (66).

$$a_{ph} = K_L*\left(\frac{\rho_L*u_L}{\mu_L^**d_p}\right)^{0.5}*\left(\frac{\rho_L*g*d_p^2}{\sigma_L}\right)^{0.45}... \tag{66}$$

The mass transfer coefficients for vapor- and liquid-side are calculated by equations (67) and (68).

$$k_{G,mersmann} = K_G * \frac{D_G}{d_p} * \frac{\varepsilon + 0.12}{\varepsilon * (1-\varepsilon)^{-1}} * \left[\frac{\rho_G * u_G * d_p}{(1-\varepsilon) * \mu_G^*} \right]^{2/3} * \left(\frac{\eta_G^*}{\rho_G * D_G} \right)^{1/3} \dots \qquad (67)$$

$$k_{L,mersmann} = K_L * \sqrt{\frac{6 * D_{L\_i}}{\pi * d_p}} * \left(\frac{\rho_L * g * d_p^2}{\sigma_L} \right)^{-0.15} * \left(\frac{u_L * g * d_p}{3} \right)^{\frac{1}{6}} \dots \qquad (68)$$

In this chapter we will emphasize the use of the Billet and Schultes correlations because they include correlations for both hydrodynamics and mass transfer. Besides, they validate their model by comparing the predicted parameters with around 3500 available experimental data points.

4.5. Structure of the Model in Aspen Custom Modeler

4.5.1. General Description

The rate-based model developed to simulate packed columns includes many sub models (or sub routines), as shown in Fig. **7**, which are linked together to the main model through the ports in the Aspen Custom Modeler (ACM®). The sub models include mass balance, convective and diffusive mass transfer in gas and liquid segments, equilibrium at the interface, and the hydrodynamic equations for pressure drop, liquid hold-up, column diameter, dimensionless parameters and packing characteristics described before. Kinetic and equilibrium reactions are included as sub-models in liquid segments and liquid bulk models.

A graphical representation of one differential element of packing is shown in Fig. **8**. Mass transfer within the gas *(GF)* and liquid *(LS)* films occurs by convection and molecular diffusion. Gradients of concentration or chemical reactions are not considered in the gas bulk *(GB)*, i.e., perfect mixing is assumed. Mass transfer occurs through the films in a direction normal to the interface *(interface)*, i.e., molecular diffusion and convection parallel to the interface are neglected. Phase equilibrium is assumed at the interface. Reactions in the liquid phase take place both in each discretized liquid segments *(LS)* and in the liquid bulk *(LB)*. The calculations in each of those sub-models (described below) uses a properties file in Aspen Properties Plus which defines the compounds, ionic species, thermodynamic method, and equilibrium reactions. As in many of the studies of absorption of CO_2 in aqueous solutions, the thermodynamic method used in this work is Elec-NRTL.

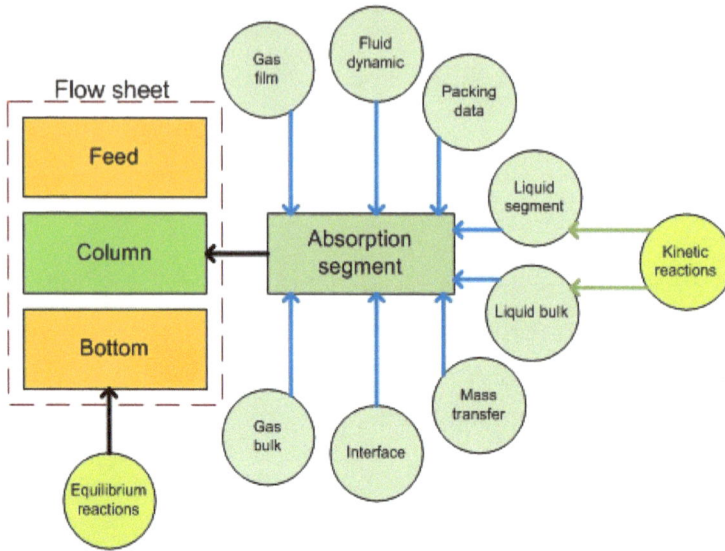

Figure 7: Model structure for chemical absorption of sour gases in a packed column.

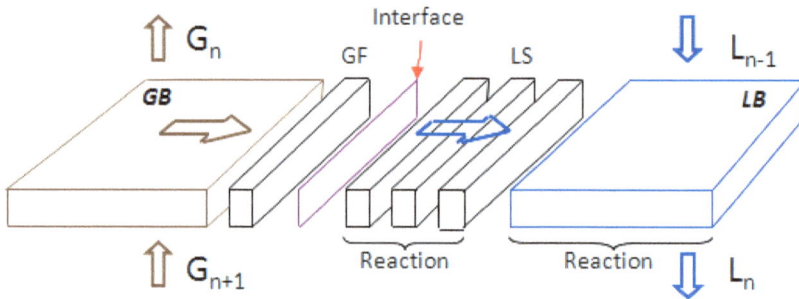

Figure 8: Graphical representation of the rate-based model with liquid phase discretization.

The code of each sub-model presented in Fig. **7** begins with the name of the sub model and finishes with the "END" sentence as presented in Fragment 1.

Fragment 1 Structure for each sub model included in the flow sheet according to ACM®

```
Model example
// <Component list name> as <ComponentListName> ("<description>");
// <parameter name> as <parameter type> (<default>,
description:"<description>");
// <variable name> as <variable type> (default, <spec>,
description:"<description>");
// <submodel name> as<model type> (<submodel variable> = <variable
```

```
name>,.);
// <structure name> as external <structure type>("<default
instance>");
// <port name> as <Input or Output><port type>;
// <equation_name>: <expression1> = <expression2>;
// Call (<output argument list>) = <procedure name>(<input
argument list>);
End
```

The previous example of a fragment of code includes declaration of: compounds, parameters, variables, ports and sub-models (if other sub-models are used to calculate some variables), equations and procedures. Ports are used to define what variables will be passed to or from a model in streams connected to the model. Additionally, the software requires de definition of type of parameters, stream type and variable types.

4.5.2. Description of Sub-Models

4.5.2.1. Feed

In the sub-model "feed", the composition of the lean solution (free-CO_2 solution) is converted into true compositions, *i.e.*, ammonia ionization in water to produce ammonium, hydronio and hydroxyl ions is considered. The true composition, molar density and pH of the lean solution are calculated by FORTRAN built-in procedures called "True_Comp", "Dens_Mol_Liq" and "pH_Value", respectively. An explanation of the input variables and the output of a procedure can be found in the library reference guide of Aspen Custom Modeler [20].

The feed sub model also calculates the volumetric flow and velocity of the lean solution and the total concentration of the solution entering at the top of the column.

4.5.2.2. Bottom

The "bottom" corresponds to a tank for collecting the rich solution (*i.e.*, having high concentration of CO_2). There, equilibrium reactions take place (the equilibrium reactions is used as sub model). In a graphical illustration (Fig. **9**) the streams and the enthalpy of the mixture are shown; the corresponding mass balance is given by equation (69), where the moles of the species *i* produced in the bottom are accounted for the term $R_{i,bottom}^{L}$. The enthalpy of the input and output streams is equal because there is no heat transfer in this section of the column and the net production of each species in the streams are zero.

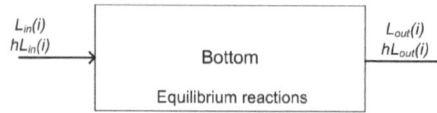

Figure 9: Graphical representation of the bottom sub model.

$$L_i^{in} - L_i^{out} + R_{i,bottom}^L = 0 \dots \tag{69}$$

4.5.2.3. Column

The model "column" is one of the most important, because it defines the ports *Stage_G_in*, *Stage_G_out*, *Stage_L_out* and *Feed* for external connections between the column and the input streams (sour gases and lean solution) and output streams (rich solution and sweet gases). This sub-model is used to connect ports and internal stages (see Fig. **10**) by the "*link*" command and for connecting internal streams by the "*connect*" command. The description of "*link*" and "*connection*" commands in the model are shown in the Fragment 2.

Fragment 2 Description of ports and connections, external and between stages in the sub model "column"

```
//***Declaration of Ports, Submodels, Connections***
Stage_G_in as Input SegmentPortGas;//Defines the port Stage_G_in for
connecting the stream steam manually during flow sheeting
Stage_G_out as Output SegmentPortGas;//Defines the port Stage_G_out for
for connecting the stream gas out manually during flow sheeting
Stage_L_out as Output SegmentPortLiquid;// the same for liquid outlet
Feed as Input SegmentPortLiquid;//The same for liquid inlet
/* *************** CONECTIONS ***************************** */
Link Stage_G_out and stage(1).G_out;//Connect the outlet gas stream of
stage 1 [stage(1).G_out] to the port Stage_G_out
Link Stage_L_out and Stage(n_stage).L_Out;//Connect the outlet liquid
stream of n-stage [Stage(n_stage).L_Out]to the port Stage_L_out
Link Stage_G_in and Stage(n_stage).G_in;//Connect the input gas stream
[Stage(n_stage).G_in] from the port Stage_G_in to stage n
Link Feed and MixStageFeed.Feed;// Connect the input liquid stream
[MixStageFeed.Feed] from the port Stage_G_in to the mixer
Connect MixStageFeed.L_out and Stage(1).L_in;//It create a stream
connecting the liquid outlet from the MixStageFeed to the liquid inlet in
Stage (1).
IF n_stage > 1 THEN
FOR i IN [2: n_stage] DO
```

```
Connect Stage(i-1).L_Out and Stage(i).L_In;//Create a stream connecting
the liquid outlet from the Stage(i-1) to the liquid inlet in Stage (i).
Connect Stage(i).G_Out and Stage(i-1).G_In;//Create a stream connecting
the gas outlet from stage (i) to the gas inlet in stage (i-1)
ENDFOR
ENDIF
```

The equations in this model define the height of each segment and calculate de the molar flow of the gas entering the column at the bottom (mol s^{-1}) by equations (70) and (71), respectively, where the volumentric flow of gas $\left(V_{Bottom}\right)$ entering at the bottom is expressed in m^3 s^{-1} and the total concentration $\left(C_{Tot}\right)$ in mol L^{-1}.

$$\Delta h = H\Big/_{n} \ldots \tag{70}$$

$$G = V_{Bottom} * C_{Tot} * 1000 \ldots \tag{71}$$

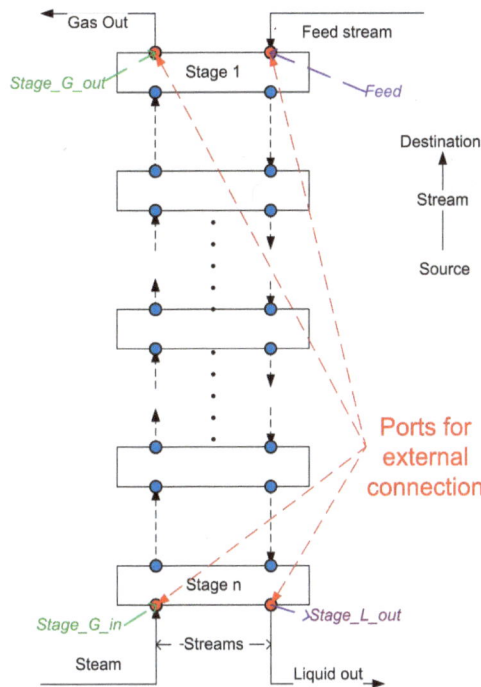

Figure 10: Sketch for the port connections in the sub model "column".

The "*connect*" command connects the gas streams (source and destination) around each packing segment (stage); the same is done for liquid streams inside the column. The number of internal connections is automatically set by a FOR cycle depending on the number of stages (packing segments) defined at the simulation start up. This model also defines the sub-model "*absorption segment*".

4.5.2.4. Absorption Segment

This model deals with the phenomena depicted in Figs. **5a** and **8**. In fact, it includes the declaration of sub-models "Gas bulk", "Gas film", "Interface", "Liquid bulk", "Packing data", "Mass transfer", and "Fluid dynamics". This model also includes the declaration of ports for connecting these sub- models, as described in Fragment 3 and shown in Fig. **11**.

Fragment 3 Sub-models and connections in the sub models Absorption segment

```
//***Declaration of Ports, Submodels
G_in as Input SegmentPortGas;//Defines the port G_in in each segment of
packing for connecting the gas streams defined in sub model "Column". See
Fragment 5.2
G_out as Output SegmentPortGas;//Defines the port G_out in each segment of
packing for connecting the gas streams defined in sub model "Column"
L_in as Input SegmentPortLiquid;//Defines the port L_out in each segment of
packing for connecting the Liquid streams defined in sub model "Column"
L_out as Output SegmentPortLiquid;//Defines the port L_out in each segment
of packing for connecting the Liquid streams defined in sub model "Column"
//***Declaration o Sub-models
Gas_Bulk as GasBulk;//includes de code for Gas bulk as a sub model of
absorption segement
Gas_Film as GasFilm;//equal
Inter_face as Interface;//equal
Liquid_Bulk as LiquidBulk;//equal
Mass_transfer as Masstransfer;//equal
Packing_data as Packings;//equal
Fluiddynamics as Fluiddynamics;//equal
//*** CONECTIONS***///
Link Gas_Bulk.GB_in and G_in;//Connect the input gas stream of gas bulk sub
model [Gas_Bulk.GB_in] to the port G_in defined for each stage
Link Gas_Bulk.GB_out and G_out;//Connect the outlet gas stream of gas bulk
[Gas_Bulk.GB_out]to the port G_out
Link Liquid_Bulk.LB_in and L_in;//Connect the input liquid stream of liquid
bulk sub model [Liquid_Bulk.LB_in] to the port L_in defined for each stage
Link Liquid_Bulk.LB_out and L_out;//Connect the outlet liquid stream of
liquid bulk [Liquid_Bulk.LB_out]to the port L_out
Connect Gas_Bulk.nGB_out and Gas_Film.nGF_in;//Create a gaseous stream
connecting the output of gas bulk to the inlet in gas film
Connect Gas_Film.nGF_out and Inter_face.nInt_in;//Create a gaseous stream
connecting the output of gas film to the inlet in the interface
Connect Inter_face.nInt_out and Liquid_Segment(1).nLS_in;//Create a stream
connecting the output of the interface to the inlet in liquid segment 1.
The liquid segment corresponds to each one of the discretized segment in
this sub model (not peresented)
```

```
If (n_seg <= 1) then//n_seg correspond to the number of segments
discretizing the liquid film
Connect Liquid_Segment(1).nLS_out and Liquid_Bulk.nLB_in;
else
For i in [2: n_seg] Do//This cycle control the conections between liquid
segments
Connect Liquid_Segment(i-1).nLS_out and Liquid_Segment(i).nLS_in;
EndFor
Connect Liquid_Segment(n_seg).nLS_out and Liquid_Bulk.nLB_in;
EndIf
```

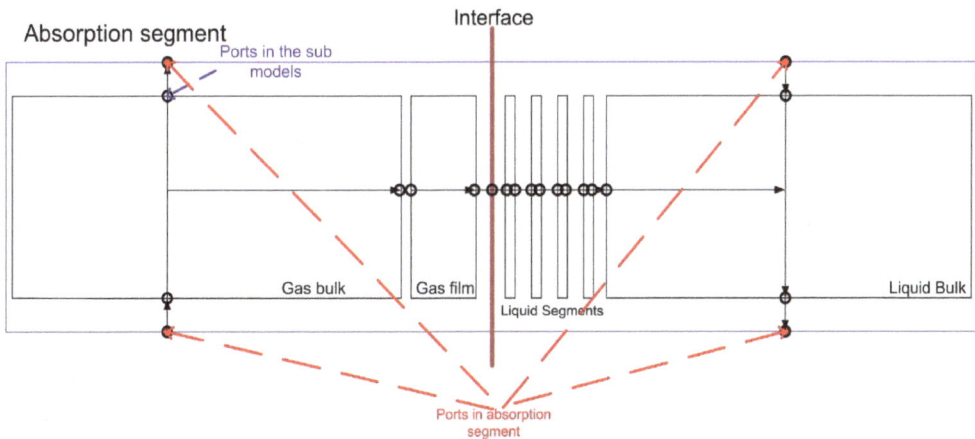

Figure 11: Connections between the ports in absorption segments and the sub-models described in the figure.

For illustration purposes, the connections between the ports in Fig. **11** were drawn as straight lines but the concentration of species may have different profiles, as will be shown later in the description of the gas film sub-model.

The liquid segments are symmetrically divided into a number of liquid sub-segments (*i.e.*, four sub-segments). Since the rate-based models are developed considering the heat transfer, different situations arise, namely, the process could be isotherm with or without heat transfer, or adiabatic. These options should be available in the model, which should also allow switching from one to the other. As an example, a simple heat transfer equation will be developed from equations (6) and (7), under the assumptions that no heat transfer toward or from the surroundings $\left(Q_j^V \text{ or } Q_j^L\right)$ are present, that is, the system is adiabatic, the side streams $\left(r_j^L + r_j^V\right)$ are equal to zero, and the change of enthalpy $\left(\Delta H_j^V \text{ or } \Delta H_j^L\right)$ is also zero because of the absence of chemical reactions. Hence, the heat transfer is solely between the gas and liquid phases. Energy balances for gas and liquid bulk in stage j are presented in equations (72) and (73). The control volumes are the gas and liquid bulk with heat transfer to or from the interface; the area for heat transfer is a_{ph}, which is evaluated by the correlations described in the section 4.4.

$$0 = G_{i,j}^{in} * h_{G\_i}^{in} - G_i^{out} * h_{G\_i}^{out} - q^V * y_i * a_{ph} * a * \Delta h \ldots \tag{72}$$

$$0 = L_{i,j}^{in} * h_{L\_i}^{in} - L_{i,j}^{out} * h_{L,i}^{out} - q^L * x_i * a_{ph} * a * \Delta h \tag{73}$$

The corresponding equations for the overall mass and energy balances without chemical reactions in a stage j are described by equations (74) and (75), respectively.

$$0 = \sum_i G_{i,j}^{in} - \sum_i G_{i,j}^{out} + \sum_i L_{i,j}^{in} - \sum_i L_{i,j}^{out} \ldots \tag{74}$$

$$0 = \sum_i G_{i,j}^{in} * h_{G\_i}^{in} - \sum_i G_{i,j}^{out} * h_{G\_i}^{out} + \sum_i L_{i,j}^{in} * h_{L\_i}^{in} - \sum_i L_{i,j}^{out} * h_{L,i}^{out} \tag{75}$$

The sub-model "absorption segments" also includes declarations of component lists, parameters, variable types, equations and procedures (FORTRAN subroutines) for calculation of enthalpy and molar mass of each compound.

4.5.2.5. Gas Bulk

Similarly to other models, the gas bulk model includes declaration of component list, variables, parameters and procedures. This model includes procedures for calculating molar density of gas, mass density and viscosity. The ports defined in this model are shown in Figs. **10** and **11**, and the mass flows in Fig. **12**.

The mass transfer, equation (76), in this sub model is derived from equation (4), where the side streams $\left(r_{j,i}^V\right)$, the inter-stage feeds $\left(f_{i,j}^V\right)$ and the net production rate of i $\left(R_{i,j}^V\right)$ in the gas bulk are equal to zero.

$$G_i^{in} = G_i^{out} + N_{i,GB} * a_{ph} * a * \Delta h \ldots \tag{76}$$

Regarding the pressure, it is assumed that there is no pressure drop (ΔP) in the gas film, in the interface, or in the liquid segment; pressure drop is calculated in the "fluid dynamic" and it is used to evaluate the pressure of the stage, equation (77).

$$P_{j+1} - P_j = \Delta P \ \ldots \tag{77}$$

The heat transfer in the gas bulk is described by equation (72), with heat transfer toward the interface through the interfacial area a_{ph}.

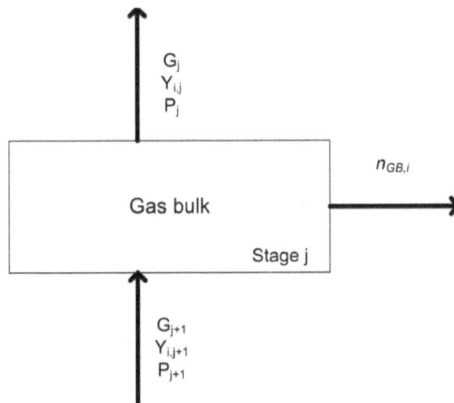

Figure 12: Representation of the streams in the sub model Gas bulk

4.5.2.6. Gas Film

The "gas film" sub-model incorporates gas-phase mass transfer limitations. The possibility of switching between different mass transfer mechanisms, *e.g.*, "Diffusion-NP", "Diffusion", and "Maxwell-Stefan" is included in the model. When the concentration of components is low, the Nernst Planck equation may be used instead of the Maxwell – Stefan. It is stated that the molar flux through the gas film is equal to the molar flux from the gas bulk and equal to the molar flux toward the interface (equation (78), that is similar to the continuity of molar flows at the interface, equation (8) in section 4.2. The molar flux of species i through the interface in stage j and relative to a stationary reference frame is described by equation (79), which is analogue to equation (10). If distillation were the subject of study, the first term in the right hand side of equation (79) would be zero due to the counter diffusion. In absorption this term is important and is expressed by equation (80) which is the mol average velocity multiplied by the concentration of species i.

$$N_i^{int} = N_i^{GB} \ldots \tag{78}$$

$$N_i^{GB} = N_i^{Tot} + J_i^{GF} \ldots \tag{79}$$

$$N_i^{Tot} = y_i^{GF} * \sum_i N_i^{GB} \ldots \tag{80}$$

$$J_i^{GF} = \frac{D_i^{GF}}{\delta x_{GF}} * \left(C_i^{GB} - C_i^{int} \right) \ldots \tag{81}$$

It is important to mention that reactions are not considered in the gas film or interface. Besides, the mass and heat transfer is only in the normal direction to the interface.

The procedures in this model are used to calculate the diffusivities of each gaseous compound, molar density and gas conductivity.

4.5.2.7. Interface

Mass transfer through the interface can be described using different theoretical approaches [21]: two-film, "still surface", and surface renewal models. Current rate-based models are based on the two-film model (see Fig. **13**) to describe the mass transfer through the interface. The main consideration of the two-film model is that mass transfer limitations arise only in the gas and liquid films at both sides of the interface. The sub-model "interface" focuses on determining the diffusion. Calculation of mass transfer coefficients is dealt with in sub-model "mass transfer".

The two-film model was adopted here because of the larger number of experimental correlations for mass transfer coefficients available for different types of packing [22]. In the interface, the two-film model uses the phase equilibrium, equation (12), where the equilibrium constant K_i is calculated by a FORTRAN sub routine called "Kvalues".

This sub-model states that the molar flows of gaseous compounds at both sides of the interface are equal and that the molar flow of electrolytic species through the interface is equal to zero.

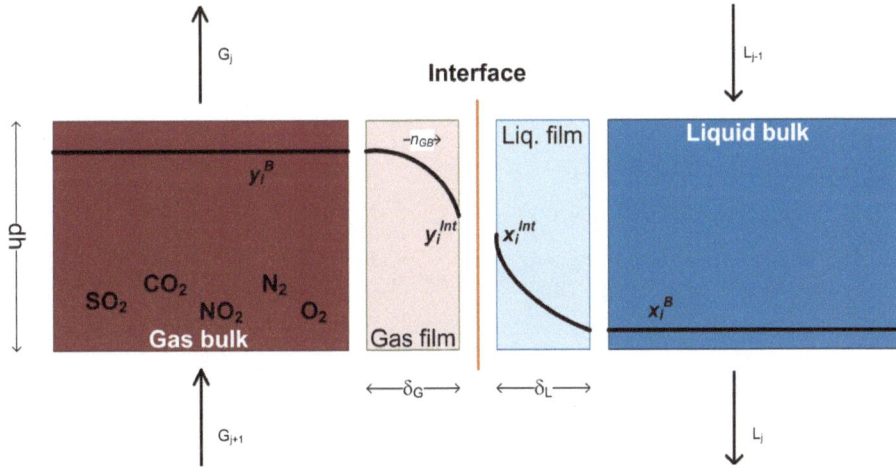

Figure 13: Description of the mass transfer simulated by the two film model in a differential segment of packing.

4.5.2.8. Liquid Segment

Similar to the gas film, the liquid segment model deals with the calculation of mass balance and mass transfer limitations for each of the sub-segments in the liquid segment discretized in the model "absorption segment". The mass balance, equation (82), in each segment with length Δx indicates that the molar flow of each species (gas o electrolyte) is equal to the inlet flow plus/minus the net production/consumption due to reactions. The net production of each species is calculated in the sub-model "reaction" which includes kinetically-controlled and equilibrium reactions. Equation (82) is the analogue to equation (5), but considering the molar flux normal to the interface $\left(N_{i,j}^{L}\right)$ and the net production of species i in each sub-segment $\left(R_{i,j}^{L}\right)$.

$$N_{i,in}^{LS} - N_{i,out}^{LS} + R_{i}^{LS} * \Delta x = 0 \qquad (82)$$

The molar flux of i through any sub-segment in the liquid segment in stage j relative to a stationary reference frame is described by equation (83), where the diffusive component is determined by equation (85), with z_i as the charge of ionic species, F_i is the Faraday constant, and the differential of volume is the product of the interfacial area and the differential length of any sub-segment $\left(a_{ph} * \delta x\right)$ in the liquid segment.

$$N_i^{LS} = N_i^{Tot,LS} + J_i^{LS} \quad \ldots \tag{83}$$

$$N_i^{Tot,LS} = x_i^{LS} * \sum_i N_i^{LS} \quad \ldots \tag{84}$$

$$J(i) = \frac{D_i^{LS}}{\delta x_{LS}} * \left(\Delta C_i^{LS}\right) + C_i^{LS} * z_i * \frac{D_i^{LS}}{\delta x_{LS}} * \frac{F_i}{R_G * T} * \delta V \quad \ldots \tag{85}$$

An important difference of these equations with those of the mass transfer in the gas segment is that the diffusive flux in the liquid segment includes a term related to the ionic species produced when acid gases such as CO_2 and SO_X are absorbed into aqueous absorbent.

Considering the shape of the concentration profile in the liquid segment shown in Fig. **13**, it is better to include as many liquid segments as possible to improve the accuracy of the simulation. However, a very large number of segments will increase the calculation time.

4.5.2.9. Liquid Bulk

The sub-model "liquid bulk" is similar to the "gas bulk": both consider perfect mixing in the corresponding phase; however, the "liquid bulk" sub-model also considers the reactions taking place in the liquid phase, which are included in the mass balance equation by the term R. The reaction term, which is the net production of each species in liquid phase (electrolyte and molecular), is calculated in the sub-model "reaction". The mass balance in the liquid bulk is derived from equation (5), but considering the side streams and inter-stage feeds equal to zero. The specific equations will be similar to equation (86), where the term h_{LS} is the liquid hold-up calculated by the correlations in section 4.4.

$$L_{i,j-1} - L_{i,j} + N_i^{LS} * a_{ph} * \left(\frac{\pi}{4} * d_{col}^2\right) * \Delta h + R_i^{LB} * \left(\frac{\pi}{4} * d_{col}^2\right) * \Delta h * h_{LS} \quad \ldots \tag{86}$$

Fig. **14** is a graphical representation of equation (86), where the contribution of the "kinetic reactions" sub-model is included.

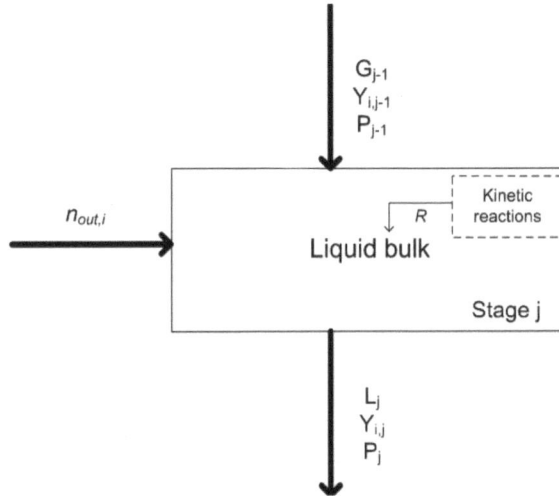

Figure 14: Representation of the streams in the sub model Liquid Bulk.

4.5.2.10. Fluid Dynamics

This sub-model uses correlations by Maćkowiak or Billet to calculate the gas capacity factor, the pressure drop and gas velocities at flooding, at pre loading and loading point. The equations included in this sub-model were already presented in the section 4.3.

4.5.2.11. Packing Data

This sub-model has the characteristic information of packing (similar to data in Tables **2** and **4**). When any type of packing is selected, all variables required for "hydrodynamics" and "mass transfer" sub-models load automatically.

4.5.2.12. Mass Transfer

This sub-model-allows choosing the desired correlation to calculate the mass transfer coefficients in the gas film and liquid segments k_G and k_L, respectively. The possibilities for mass transfer correlations are: "Maćkowiak", "Billet/Schultes", "Onda" and "Zech/Mersmann", which were already detailed in section 4.4.

4.5.2.13. Kinetics and Equilibrium Reactions

For describing the option of including reactions in a RBM, let's consider the case where CO_2 and SO_2, previously dissolved from the gas bulk, react in the liquid phase (aqueous ammonia solution) by a reaction network which includes kinetic and equilibrium reactions, as described by equations (87) to (93), where equation (87) and (88) are kinetically-controlled reactions. The forward reaction rate constant for equation (87) is listed in Table **6** [23] and the reaction rate constant for equation (88) is shown below; the reverse reaction rate constants are derived from the equilibrium expressions listed in Table **7**.

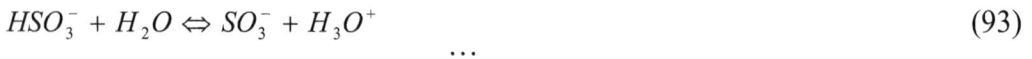

$$CO_2 + OH^- \Leftrightarrow HCO_3^- \quad \cdots \tag{87}$$

$$CO_2 + NH_3 + H_2O \Leftrightarrow NH_2COO^- + H_3O^+ \quad \cdots \tag{88}$$

$$NH_3 + H_2O \Leftrightarrow NH_4^+ + OH^- \quad \cdots \tag{89}$$

$$2H_2O \Leftrightarrow H_3O^+ + OH^- \quad \cdots \tag{90}$$

$$HCO_3^- + H_2O \Leftrightarrow CO_3^{2-} + H_3O^+ \quad \cdots \tag{91}$$

$$SO_2 + 2H_2O \Leftrightarrow HSO_3^- + H_3O^+ \quad \cdots \tag{92}$$

$$HSO_3^- + H_2O \Leftrightarrow SO_3^- + H_3O^+ \quad \cdots \tag{93}$$

Table 6: Parameters for forward reaction constants of kinetic controlling reaction

Kinetic Controlling Reactions $k_i = 10^{(\alpha_i - \beta_i / T)}$		
Reaction	α	β
Equation (87)	13.635	2,895

Equations (89) to (93) are considered to attain equilibrium instantaneously. Reactions of SO$_2$ with ammonia were not included because of the relatively low concentration of SO$_2$ with respect to that of CO$_2$. The solubility equilibrium corresponding to the system CO$_2$-NH$_3$-H$_2$O were not considered because the lean solution is fed as CO$_2$-free, the relative high dilution of the lean solution (low concentration of ammonia), and the short contact between aqueous ammonia and CO$_2$. These conditions guarantee a concentration of ions in the solution below the saturation, avoiding the precipitation of solid species such as bicarbonates, carbonates and carbamates.

Table 7: Parameters for determining equilibrium constants of reaction

Equilibrium Reactions	$In(K) = A + B/T + C * In(T) + d * T$				
Reaction/Equation	A	B	C	D	T(°C)
(87)	-38.6565	5,719.89	7.97117	0.0279842	0-100
(88)	226.882	-9,192.1	-36.7816	0	0-100
(89)	2.76	-3,335.7	1.4971	0.0370566	0-100
(90)	140.932	-13,445.9	-22.4773	0	0-100
(91)	220.067	-12431.7	-35.4819	0	0-100
(92)	-5.978673	637.396	0	-0.0151337	0-100
(93)	-25.290564	13333.4	0	0	0-100

Reaction corresponding to equation (88) is the main contributing reaction to CO_2 absorption. In previous studies conducted by the authors it was found that the reaction takes place only in the liquid phase and its behavior was explained by the Zwitterion mechanism (equation (94), with specific constants summarized in Table **8**.

$$r_{CO_2-NH_3} = \frac{C_{CO_2}C_{NH_3}}{\left(\frac{1}{k_2}\right) + \left(\frac{1}{k_{NH_3}C_{NH_3} + k_{H_2O}C_{H_2O}}\right)} = k_{app} * C_{CO_2} \dots \tag{94}$$

Table 8: Specific constants for Zwitterion equation

T [K]	k_2	k_{NH_3}	k_{H_2O}
278.15	19,000	353.4	13.3
284.65	20,950	440.0	18.9
291.15	24,298	556.4	29.2
299.15	27,027	737.8	48.6
308.15	28,500	883.5	57.0

The reactions sub models ("equilibrium reactions and kinetic reactions") consider the non-ideal behavior of the liquid phase by including activity coefficients in the equilibrium constant of each reaction, equation (95), and the temperature-dependence of the equilibrium constant, equation (96); the activity coefficients are calculated by procedures in the ACM®. The constants for kinetically-controlling reactions for the absorption of CO_2 with ammonia are described by equations (97) and (98). The net rate of production of the species i in the liquid segment of stage j is given by the equation (99), with m and n as equilibrium and kinetic reactions. The subscript c and p denotes consumption and production of species i.

$$K_m = \prod_p (\gamma_p * x_p)^{v_p} \Big/ \prod_r (\gamma_r * x_r)^{v_r} \quad \ldots \tag{95}$$

$$\ln(K_m) = A + \frac{B}{T} C * \ln(T) + d * T \quad \ldots \tag{96}$$

$$k_Z = f\left(k_2, k_{NH_3}, k_{H_2O} C_{NH_3}\right) \ldots \tag{97}$$

$$k_n = 10^{\left(\alpha - \frac{\beta}{T}\right)} \ldots \tag{98}$$

$$R_{i,j}^L = \sum_m r_{i,p}^m - \sum_m r_{i,c}^m + \sum_n r_{i,p}^n - \sum_n r_{i,c}^n \quad \ldots \tag{99}$$

4.6. Thermodynamic Model

Since the content of this chapter is concerned with developing a model for absorption (reactive absorption) of CO_2, which is frequently an electrolyte system, the main model described here is the Electrolyte-NRTL activity coefficient model. ELECNRTL calculates the liquid phase properties from the electrolyte-NRTL activity coefficient model. Vapor phase properties are calculated from the Redlich-Kwong equation of state. The ELECNRTL can represent aqueous and aqueous/organic electrolyte systems over the entire range of electrolyte concentrations with a single set of binary interaction parameters. In the absence of electrolytes, the model reduces to the standard NRTL model. The ACM® software uses interaction parameters from a definition file of Aspen Properties®.

Since two types of ideality have been defined, one leading to a Raoult's law and the other to the Henry's law, activity coefficients may be normalized in two different ways:

If the activity coefficients are defined with reference to an ideal solution according to Raoult's law, normalization for each component i is expressed by equation (100).

$$\gamma_i \to 1 \quad when \quad x_i \to 1 \ldots \tag{100}$$

Since this normalization applies both for solute and solvent, it is called a symmetric criterion for normalization.

However, if the activity coefficients are defined with reference to an ideal dilute solution (*e.g.*, for electrolytic solutions), we have, equation (101):

$$\begin{aligned} \gamma_i \to 1 \quad when \quad x_i \to 1 \quad (solvent) \\ \gamma_m^* \to 1 \quad when \quad x_m \to 0 \quad (solute) \end{aligned} \ldots \tag{101}$$

As the solute and solvent are not normalized in the same way, equation (101) is an asymmetric criterion of normalization [24]. In order to distinguish between symmetric and asymmetric activity coefficients, in the latter the activity coefficient of a component is marked with an asterisk (*).

4.7. The Electrolyte NRTL Model

The Electrolyte NRTL model was originally proposed by Chen *et al.*, 1982, 1986 [25, 26] for aqueous electrolyte systems. It was later extended to mixed solvent electrolyte systems by Mock *et al.*, 1984 and 1986 [27, 28]. The model is based on two fundamental assumptions:

The like-ion repulsion assumption: states that the local composition of cations (or anions) around cations (or anions) is zero. This is supported on the assumption that the repulsive forces between ions having the same charge are extremely large.

The local electro-neutrality assumption: states that the distribution of cations and anions around a central molecular species is such that the net local ionic charge is zero.

In the excess Gibbs energy expression proposed by Chen, two contributions were included: one for the long-range ion-ion interactions (represented by the asymmetric Pitzer-Debye-Hückel (PDH) model and the Born equation) that exist beyond the immediate neighborhood of a central ionic species, and the other related to the local interactions (represented by the Non-Random Two Liquid (NRTL) theory) that exist at the immediate neighborhood of any central species. The NRTL expression for the local interactions, the Pitzer-Debye-Hückel expression, and the Born's equation are combined to give equation (102) for the excess Gibbs energy.

$$\frac{G^{E*}}{RT} = \frac{G^{E*,PHD}}{RT} + \frac{G^{E*,Born}}{RT} + \frac{G^{E*,NRTL}}{RT} \cdots \tag{102}$$

Where the asterisk (*) denotes an asymmetric reference state, well-accepted in electrolyte thermodynamics, R is the ideal gas constant, and T is the absolute temperature. The asymmetric activity coefficients γ_i^* are obtained from the partial differentiation of G^{E*} (for PDH model and Born's equation) with respect to the number of mole of species i, n_i. Hence, the expression for the asymmetric activity coefficient is derived by replacing the differential of G^{E*}, giving equation (103):

$$\ln\left(\gamma_i^*\right) = \ln\left(\gamma_i^{*,PDH}\right) + \ln\left(\gamma_i^{*,Born}\right) + \ln\left(\gamma_i^{*,NRTL}\right) \tag{103}$$
$$\cdots$$

The Pitzer-Debye-Hückel (PDH) activity coefficient is used to represent the long-range interaction contribution. The Debye-Hückel theory is based on the infinite dilution reference state for ionic species in the actual solvent media. For systems with water as the only solvent, the reference state is the aqueous solution at infinite dilution. For mixed-solvent systems, the reference state for which the Debye-Hückel theory remains valid is the solution with the corresponding mixed-solvent composition at infinite dilution.

The Born excess Gibbs energy is used to account for the Gibbs energy of transfer of ionic species from the infinite dilution state in a mixed-solvent to the infinite dilution state in aqueous phase.

The local interaction contribution is developed as a symmetric model, based on reference states of pure solvent and pure completely dissociated liquid electrolyte, and the asymmetric model is obtained by normalization of symmetric model

$\left(G^{E,NRTL}/RT\right)$ with infinite dilution activity coefficient $\left(\gamma_m^{\infty,NRTL}/RT\right)$ as shown in equation (104).

$$\frac{G^{E*,NRTL}}{RT} = \frac{G^{E,NRTL}}{RT} - \sum_m x_m \ln\gamma_m^{\infty,NRTL} \tag{104}$$

4.8. Cubic Equations of State

4.8.1. The SKR (Soave-Redlich-Kwong) Equation of State

Equation (105) corresponds to the pressure explicit form of SKR equation [29]:

$$P = \frac{RT}{V-b} - \frac{a}{V(V+b)} \tag{105}$$

Where P is the system pressure, V is the molar volume, T is the absolute temperature, and R is the ideal gas constant. The parameters a and b are given by temperature dependent expressions according to equations (106) and (107), which are a modification of the ones proposed by Redlich and Kwong [30].

$$a_i = 0.42748\alpha_i \frac{R_2 T_{c,i}^2}{P_{c,i}} \tag{106}$$

$$b_i = 0.08664 \frac{RT_{c,i}}{P_{c,i}} \tag{107}$$

The expression for the fugacity coefficient φ_i for the SKR equation of state is described by equation (108).

$$\ln\varphi_i = \frac{b_i}{b}(Z-1) - \ln\left(Z - \frac{Pb}{RT}\right)$$
$$- \frac{a}{bRT}\left(\frac{2}{a}\sum_k x_k a_{jk}(1-k_{ik}) - \frac{b_i}{b}\right)\ln\left(1 + \frac{b}{v}\right) \cdots \tag{108}$$

Where Z is the compressibility factor, which is represented by equation (109).

$$Z = \frac{PV}{RT}$$ (109)

...

5. SIMULATION OF THE CO$_2$ ABSORPTION WITH A RBM

The RBM developed in section 4 of this chapter is tested by simulating the simultaneous absorption of CO$_2$, SO$_2$ and NO$_2$ in a packed tower. An absorption target of 80% for CO$_2$ was set while maintaining ammonia slip below 10 ppm. The absorption was carried out in a column with 1.0 m diameter and 3 m height. These values may be modified if the F_V and absorption efficiency of CO$_2$ does not meet 1.0 Pa$^{0.5}$ and 80%, respectively. Conditions for simulation are described in Fig. **15**, where 10 stages have been defined for the simulations.

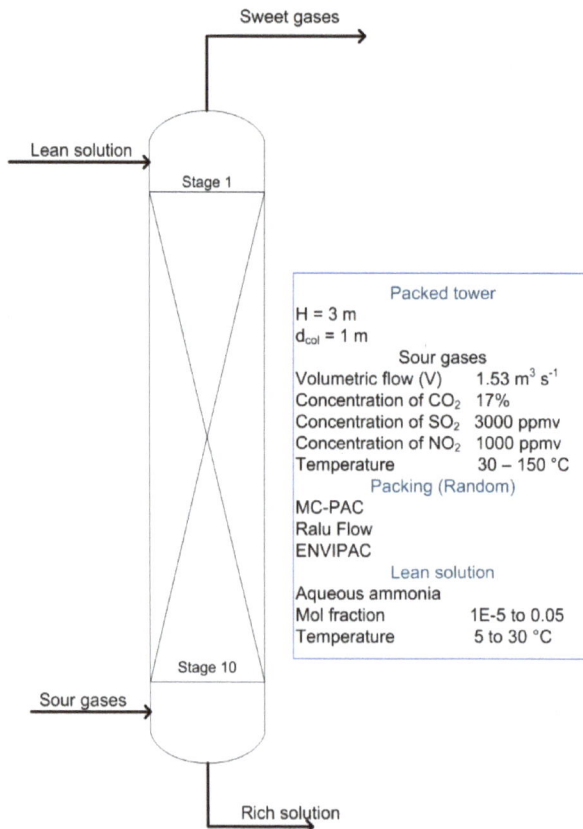

Figure 15: Conditions of simulation for absorption of CO$_2$, SO$_2$ and NO$_2$ in a packed tower by using a RBM.

Mass transfer coefficients in the model were evaluated by Billet's correlations (see section 4.4.1) and the hydrodynamics by Maćkowoak's correlations (see Table **5** in section 4.3). The reactions (equilibrium and kinetically-controlled) are those described in section 4.5.2.13. Different variables affecting the absorption efficiency were evaluated, namely temperature of the lean solution and sour gases, and concentration of ammonia in the lean solution.

The effect of streams temperature and ammonia concentration in lean solution over the abatement efficiency of CO₂ and ammonia slip.

When the temperature of the lean solution decreased from 30 to 5 °C, the ammonia slip decreased from around 4,500 to 1,357 ppm (Fig. **16a**). On the other hand, when the gas inlet temperature decreased from 150 to 30 °C, the ammonia slip decreased from 6,960 to 6,870 ppm (Fig. **16b**). Ammonia slip was significantly higher than 10 ppm in both cases. Furthermore, it is observed that ammonia slip depends more strongly on the temperature of the lean solution than on that of the sour gases entering the column. Data on Figs. **16a** and **16b** were obtained using the same flow rates of sour gases and lean solution entering the column, but different molar fraction of ammonia in the lean solution (0.004 in Fig. **16a** and 0.006 in Fig. **16b**), which explains the larger ammonia slip in Fig. **16b**. When the lean solution and the sour gases enter at 30 and 150 °C, respectively, the highest ammonia slip are 4,500 and 6,960 ppm for the lean solutions with molar fraction of ammonia 0.004 and 0.006, respectively, which suggests a dependence of the ammonia slip with ammonia concentration in the lean solution.

The inlet gas temperature affects the absorption efficiency, as shown in Fig. **17**, absorption efficiency at 150 °C is 20% higher than that at 60 °C. Although results of Fig. **17** suggest operating at high temperature, it is preferable to carry the process at low temperature because this condition favors a greater heat recovery from the effluent stream, and reduces the flow rate of gases entering the column and the ammonia slip.

Similarly to the results obtained by varying the gas inlet temperature, the efficiency of CO₂ absorption increased with the temperature of the lean solution, Fig. **18**. In fact, efficiency of CO₂ absorption is 15% higher when the temperature of the lean solution is increased from 5 to 29 °C. However, it would be more convenient to absorb CO$_2$ at 5 °C due to the considerably lower ammonia slip, Fig. **16a**.

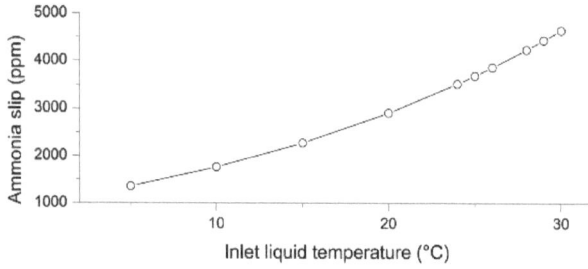

a)

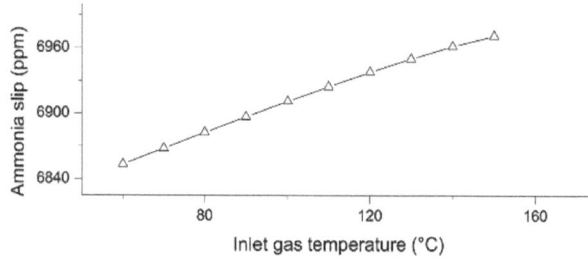

b)

Figure 16: Effect of streams temperature on ammonia slip: a) Effect of inlet liquid temperature when the mole fraction of ammonia in lean solution was 0.004 and sour gases entering at 150 °C and b) Effect of gas inlet temperature when lean solution at 30 °C was used and having 0.006 mole fraction of ammonia.

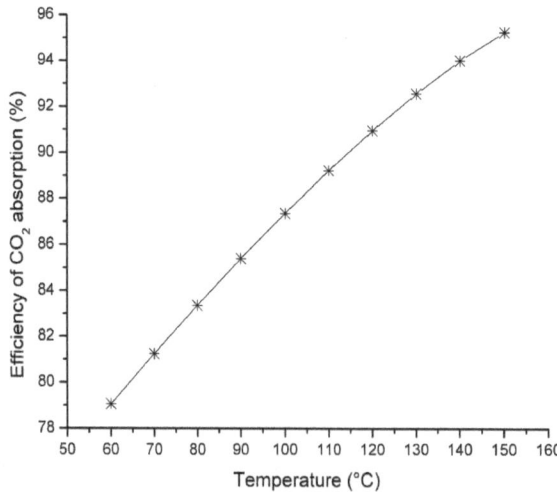

Figure 17: Effect of gas inlet temperature over efficiency of CO$_2$ absorption.

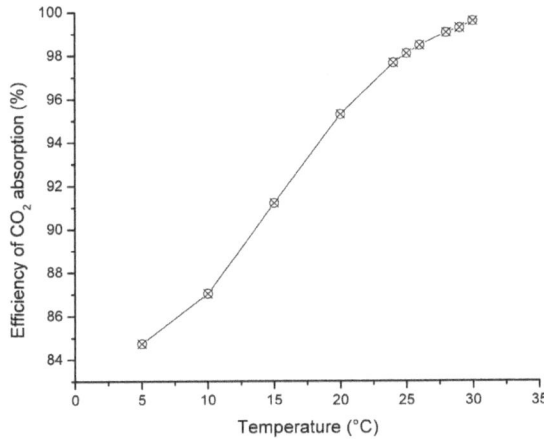

Figure 18: Effect of liquid feed temperature over efficiency of CO_2 absorption.

According to Figs. **16**, **17** and **18**, ammonia slip and absorption efficiency for CO_2 decrease if the temperature of the inlet liquid and gas streams is lowered. However, the lowest ammonia slip (near 1,400 ppm) reached in Fig. **16a** (0.004 molar fraction of ammonia) remains above the 10 ppm limit. As a consequence, the concentration of the lean solution required to meet the desired limit for ammonia slip was determined. The simulation results indicate that ammonia slip was drastically reduced from 1,357 to 3.15 ppm (see Fig. **19**) when the molar fraction of ammonia in the lean solution decreased from 0.004 to 0.00001. Notwithstanding, CO_2 absorption efficiency dropped from 85 to 61%. Therefore, the flow rate of the lean solution had to be increased by 24% to obtain CO_2 absorption larger than 80%.

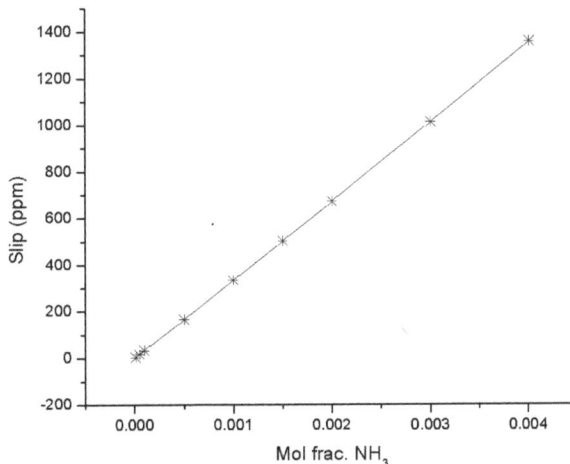

Figure 19: Effect of ammonia concentration on ammonia slip.

When the mol fraction of the lean solution was reduced to 0.00001, it was found that a flow rate of 2,100 m^3 h^{-1} of lean solution at 5 °C was required. Under these conditions, the ammonia slip was 3.2 ppm, CO$_2$ absorption efficiency was 83%, and the gas capacity factor (F_V) at the top of the column was 1.1 Pa$^{0.5}$. These results indicate that a single column, Fig. **20a**, would be sufficient to absorb more than 80% CO$_2$ with an acceptable F_V. Even though absorption of CO$_2$ using a high ammonia concentration in the lean solution would increase the absorption efficiency and reduce the liquid load and, hence, column diameter, the ammonia slip would be increased as in Fig. **19**. Therefore, as shown in Fig. **20b**, a second packed column would be required to absorb the ammonia released. For instance, when the molar fraction of ammonia is increased from 1E-05 to 0.05, ammonia slip in the first column in Fig. **20b** increases from 3.2 ppm to 24,233 ppm while the required flow of lean solution to achieve a 80% CO$_2$ absorption efficiency is reduced by to 200 m^3 h^{-1}; the gas capacity factor at these operating conditions is 1.2 Pa$^{0.5}$. The dimensions of the second absorber in Fig. **20b**, used to ensure an ammonia slip below 10 ppm, were 0.6 m of diameter and 3 m height; the column was packed with Mc-Pac, and water (2.9 m^3 h^{-1} at 12 °C) was used as the absorbent. With this arrangement, ammonia slip in stream 5 in Fig. **20b** was 7.4 ppm. The gas capacity factor (F_V) of absorber 02 was 3.7 Pa$^{0.5}$. Absorber 02 was also modeled with a rate-based model, but without including any liquid-phase reactions. The results from the simulation of schemes shown in Fig. **20** are summarized in Table **9**; the gas flow entering the absorber 01 is the same for both schemes.

Profiles of molar flow rate for CO$_2$, SO$_2$, NO$_2$ and NH$_3$ in the gas stream along absorbers 01 (Figs. **20a** and **20b**) are shown in Fig. **21**. Results confirm that SO$_2$ and NO$_2$ are quickly absorbed at the bottom of the column (Figs. **21b** and **21d**), whereas absorbing 80% of the CO$_2$ requires the entire column (Fig. **21a**). This is a result of the large CO$_2$ concentration in the gas stream and its low solubility, compared to SO$_2$ and NO$_2$. Molar flow of gaseous ammonia along the absorber 01 in Fig. **20a** is lower than that along absorber 01 in Fig. **20b** (see Fig. **21c**) due to the higher concentration of ammonia in the lean solution entering absorber 01 in the schema of Fig. **20b**.

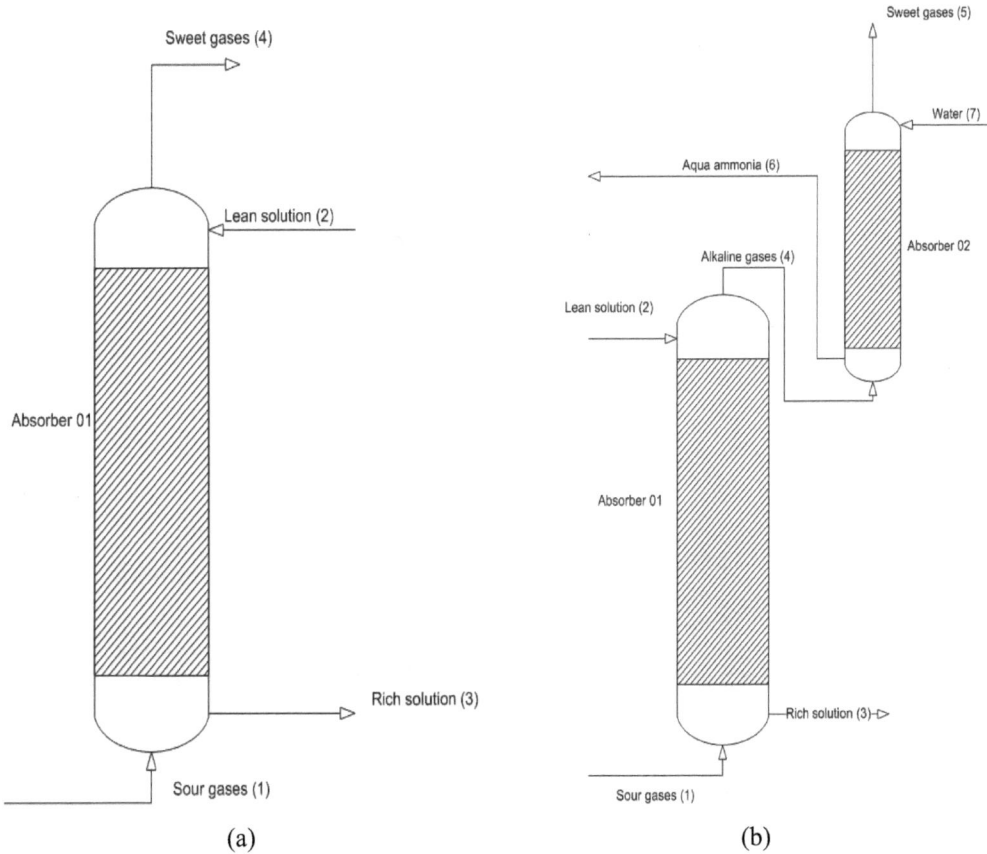

(a)   (b)

Figure 20: Schemas for CO_2, SO_2 and NO_2 absorption from a gaseous stream; a) using a diluted solution of ammonia (1.0E-05 molar fraction) at 5 °C and b) using a concentrated solution of ammonia (5.0E-02), a second tower is required to kept ammonia slip below 10 ppm.

Table 9: Simulation results of a single packed column (Fig. **20a**) and two packed columns (Fig. **20b**)

Stream	Molar Flow [kmol h-1] (Fig. 20a)	Molar Flow [kmol h-1] (Fig. 20b)
Sour gases (1)	209.9 (5500 Nm3 h-1)	209.9 (5500 Nm3 h-1)
CO2	35.7	35.7
SO2	1.7	1.7
N2	125	125
O2	20.8	20.8
NO2	0.4	0.4
H2O	26.3	26.3
Lean solution (2)	116570.6 (2100 m3 h-1)	10819.6 (200 m3 h-1)
Water	116569.0	10249.2
NH3	1.6	570.4

Table 9: contd...

Rich solution (3)	116649.4	10855.9
Sweet gases (4)	130.6	146.4
CO2	6.0	7.2
SO2	2.1E-12	4.6E-15
N2	123.5	124.8
O2	6.3E-07	9.9
NO2	5.5E-14	4.6E-14
H2O	1.1	1.2
NH3	4.2E-04 (3.2 ppmv)	3.3 (24232.6 ppmv)
Sweet gases (5)		145.6
CO2		7.2
SO2		4.0E-15
N2	N/A	124.8
O2		8.5
NO2		4.4E-10
H2O		5.1
NH3		4.6E-04
Aqueous ammonia (6)	N/A	161.8
Water (7)	N/A	161

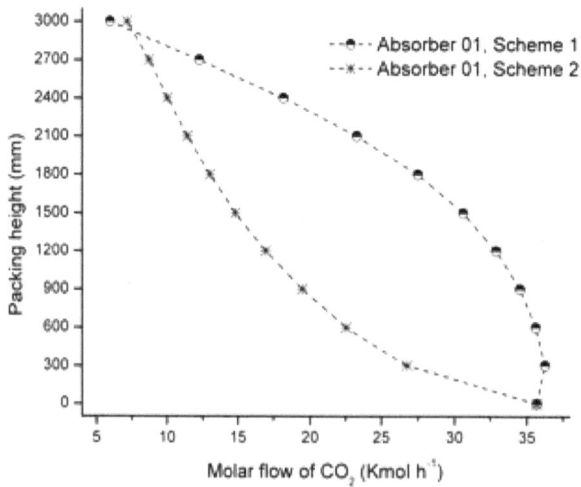

a)

Fig. 21: contd...

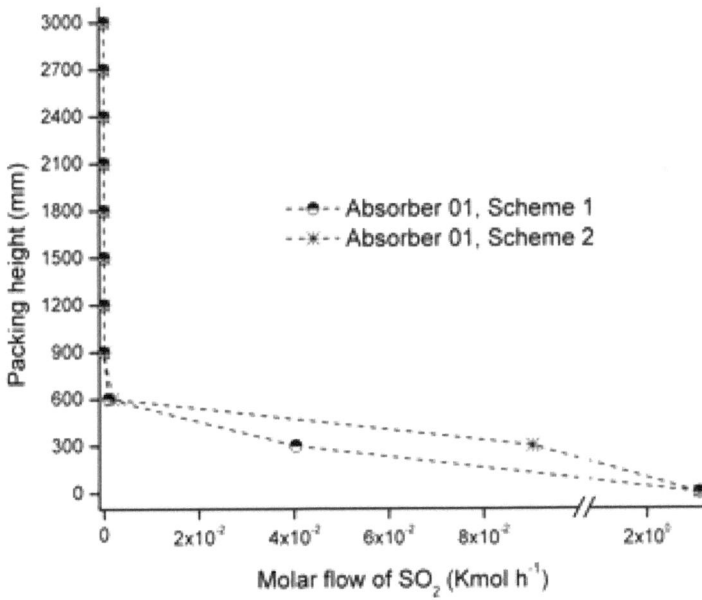

b)

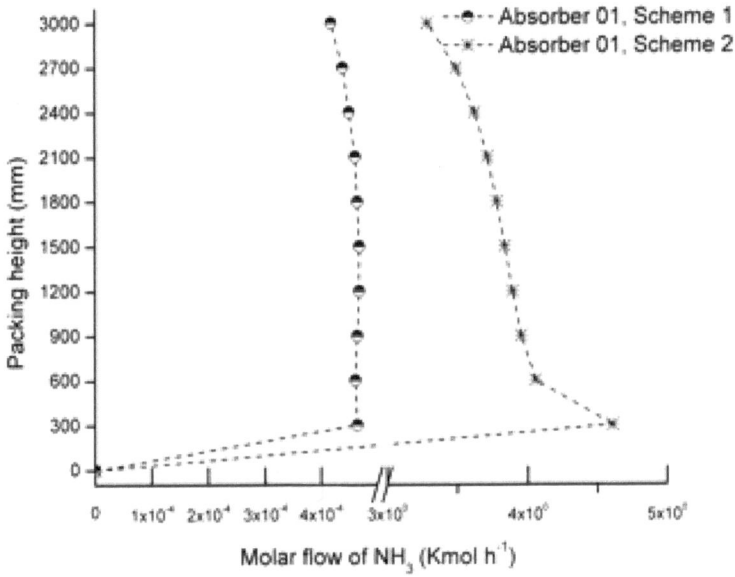

c)

Fig. 21: contd...

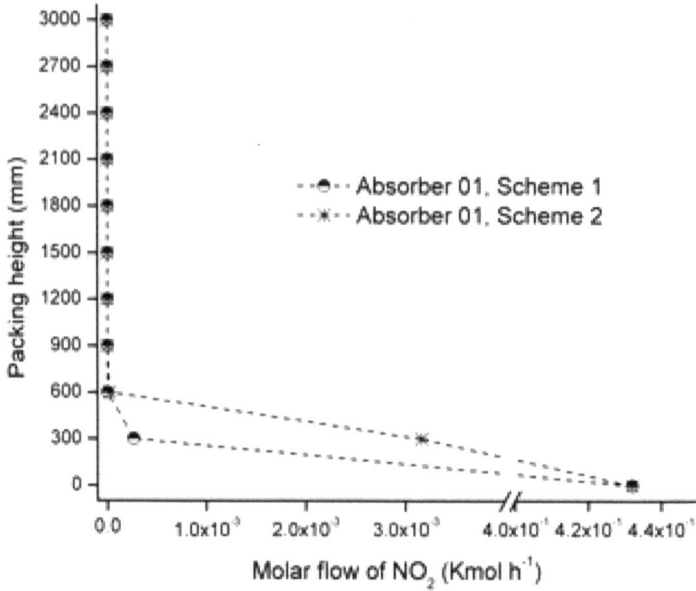

d)

Figure 21: Molar gas flows of CO_2 (a), SO_2 (b), NH_3 (c) and NO_2 (d) along absorber 01 in Schemes 1 (Fig. **20a**) and Scheme 2 (Fig. **20b**).

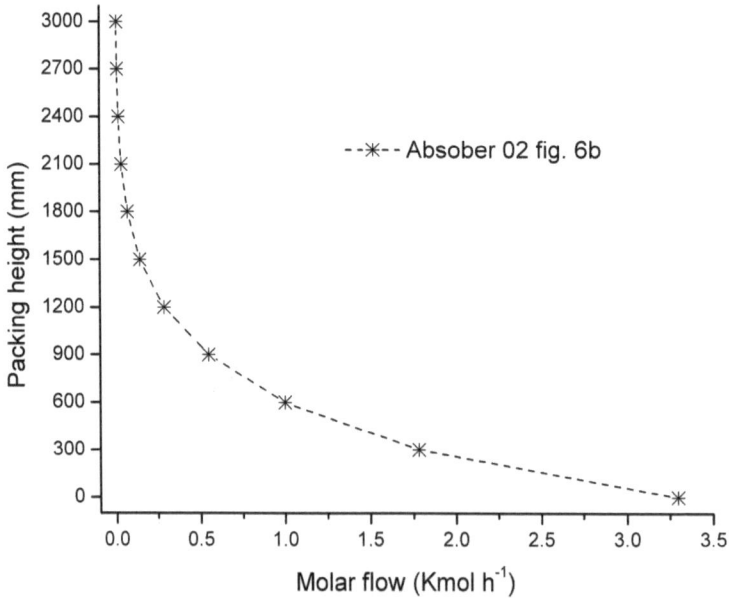

Figure 22: Molar gas flow of ammonia along absorber 02 (washer).

Table 10: Mole fractions of main gases along absorbers 01 and absorber 02 for both schemes in Fig. **20**

Height of the Column (mm)	Absorber 01 Schema 1				Absorber 01 Schema 2				Absorber 02 Schema 2			
	CO$_2$	SO$_2$	NO$_2$	NH$_3$	CO$_2$	SO$_2$	NO$_2$	NH$_3$	CO$_2$	SO$_2$	NO$_2$	NH$_3$
3000	4.6E-02	1.6E-14	4.2E-16	3.2E-06	4.9E-02	3.2E-17	3.2E-16	2.2E-02	5.0E-02	3.1E-17	6.5E-16	1.1E-05
2700	8.8E-02	2.0E-14	4.2E-16	3.2E-06	5.6E-02	1.0E-16	3.2E-16	2.2E-02	5.0E-02	3.2E-17	6.7E-16	3.7E-05
2400	1.3E-01	2.6E-14	4.2E-16	3.1E-06	6.2E-02	2.3E-16	3.2E-16	2.2E-02	5.0E-02	3.2E-17	7.0E-16	9.2E-05
2100	1.6E-01	4.0E-14	4.2E-16	3.1E-06	6.9E-02	4.6E-16	3.3E-16	2.2E-02	5.0E-02	3.2E-17	7.3E-16	2.1E-04
1800	1.8E-01	7.4E-13	4.2E-16	3.0E-06	7.8E-02	2.4E-14	4.3E-16	2.3E-02	5.0E-02	3.2E-17	7.5E-16	4.6E-04
1500	1.9E-01	4.4E-11	4.2E-16	3.0E-06	8.7E-02	6.4E-12	1.9E-14	2.3E-02	5.0E-02	3.2E-17	7.6E-16	9.7E-04
1200	2.1E-01	2.4E-09	7.5E-16	2.9E-06	9.9E-02	1.1E-09	3.5E-12	2.3E-02	5.0E-02	3.2E-17	7.5E-16	1.9E-03
900	2.1E-01	1.2E-07	5.6E-13	2.9E-06	1.1E-01	1.3E-07	6.2E-10	2.3E-02	5.0E-02	3.2E-17	6.9E-16	3.7E-03
600	2.2E-01	5.6E-06	9.5E-10	2.8E-06	1.3E-01	1.0E-05	1.1E-07	2.3E-02	4.9E-02	3.2E-17	5.8E-16	6.8E-03
300	2.2E-01	2.4E-04	1.6E-06	2.8E-06	1.5E-01	5.0E-04	1.7E-05	2.5E-02	4.9E-02	3.2E-17	4.4E-16	1.2E-02
0	1.7E-01	8.0E-03	2.0E-03	0.0E+00	1.7E-01	8.0E-03	2.0E-03	0.0E+00	4.9E-02	3.2E-17	3.2E-16	2.2E-02

The ammonia slip in Fig. **20** is reduced from 24,232.6 to 7.4 ppm, by adding the washer column, *i.e.*, absorber 02. The profile of molar flow rate of ammonia along the washer is shown in Fig. **22**. Gases from absorber 01 in Fig. **20b** enter at the bottom of absorber 02 (at around 5.5 °C), whereas water at 12 °C enters at the top of the washer. As can be observed in Fig. **22**, despite the high solubility of ammonia in water, almost the whole length of the column (7 of 10 stages) is required to meet ammonia slip lower than 10 ppm. Data of gas composition along absorbers 01 and 02 (Fig. **20b**) are listed in Table **10**.

6. SCALING AN ABSORPTION TOWER FOR REAL APPLICATIONS FROM SIMULATION RESULTS

In order to design a large-scale column, flow rates of the lean solution, sour gases and their ratio at the column inlet have to be specified. Besides the flow ratio, temperatures of sour gases and lean solution, ammonia composition, and transport properties of gas and liquid phases are also required. These values were taken from simulation results presented in section 5 and combined with the flow of gases from an actual process.

Applying the phase flow ratio (λ_0), the flow rate of lean solution to absorb 80% of the CO$_2$ present in an actual gas stream (57,864 m^3 h^{-1}, 1 bar, 30 °C) would be 6,840 m^3 h^{-1}, if the schema in Fig. **20a** were adopted, whereas 2,105 m^3 h^{-1} would be required with schema in Fig. **20b**.

The required inputs for sizing the column are listed in Table **11** and the characteristics of selected packing (Mc-Pac, Envipac and Ralu-flow) are listed in Table **2**.

Table 11: Flows and properties used to size absorbers 01 and 02 according to Fig. **20a** and **20b**

Variable	Schema in Figure 20a	Schema in Figure 20b	
	Absorber 01	Absorber 01	Absorber 02
Inlet gas flow rate (m^3 h^{-1})	57864	57864	34635
Inlet liquid flow rate (m^3 h^{-1})	6839	2105	30.5
Density of gas (kg m^{-3})	1.260	1.262	1.163
Density of liquid (kg m^{-3})	988.89	971.31	999.522
Liquid surface tension (N m^{-1})	0.0758	0.0741	0.0754
Liquid viscosity at column top (kg m^{-1} s^{-1})	1.06E-03	9.89 E-04	9.62E-04
Gas viscosity at column top (kg m^{-1} s^{-1})	1.66E-05	1.67E-05	1.71E-05
Relative column load	0.8	0.8	0.8

Afterwards, liquid hold-up, gas velocity at flooding, column diameter, liquid hold-up and pressure drop at loading point were calculated according to the iterative process described in Fig. **23** and using equations included in section 4.3. The hydrodynamic results and the size of the columns are presented in Table **12**, where different packing materials were used in the simulation of schema in Fig. **20b**.

Table 12: Hydrodynamic results of the large scale absorber 01 for schemes considered Fig. **20**

Issue	Figure 20a	Figure 20b					
	Absorber 01	Absorber 01			Absorber 02		
	Mc-Pac	Mc-Pac	Envipac	Ralu – flow	Mc-Pac	Envipac	Ralu – flow
λ_0	9.46E-02	2.91E-02	2.91E-02	2.91E-02			
$h^0_{L,FL}$	4,74E-01	3.08E-01	3.08E-01	3.08E-01	5.73E-02	5.73E-02	5.73E-02
$u_{G,FL}$	4.55E-01	1.24	1.36	1.35	3.70E00	4.26	4.23
d_s	7.5	4.5	4.3	4.4	2.0	1.9	1.9
u_G	3.64E-01	9.93E-01	1.09	1.08	2.96E00	3.41	3.38
u_L	4.30E-02	3.61E-02	3.95E-02	3.92E-02	2.61E-03	3.00E-03	2.98E-03
$F_{G,FL}$	5.11E-01	1.39	1.52	1.51	3.99E00	4.60E00	4.56E00
F_G	4.09E-01	1.12	1.22	1.21	3.19E00	3.68E00	3.65E00
$\Delta P_0/H$	5.5	30.6	22.5	23.9	294.7	196.8	208.3
$\Delta P/H$	20.4	94.9	74.8	77.3	343.3	233.3	230.2
$\Delta P/H_{(0.65 <Fr< 1.0)}$	29.0	145.9	114.8	120.4	436.3	292.9	289.1
$(\Delta P_{FL}/H)$	87.2	525.4	403.8	456.6	980.5	624.5	700.0
h_L	0.133	0.131	0.134	0.141	0.024	0.024	0.025
$h_{L,S}$	0.169	0.148	0.151	0.157	0.027	0.027	0.028

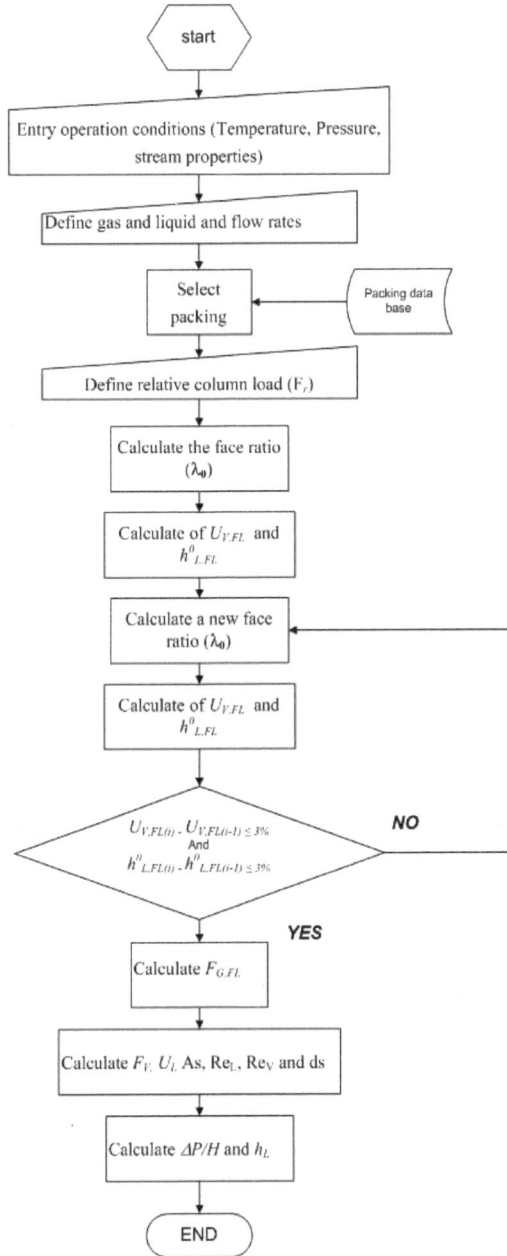

Figure 23: Iterative method for determining the diameter of the column.

Results indicate that the higher flow of lean solution entering absorber 01 (Fig. **20a**) results in the requirement of larger column diameter (7.5 m) than in the schema in Fig. **20b** (4.5 m). Moreover, absorber 01 of Fig. **20a** has a very low pressure drop and the gas capacity factor was lower than 1.0. This will result in lower efficiency and more difficult column operation. In fact, when the capacity factor is lower than 1.0 and the liquid load is high, the flooding point may be reached by a slight variation in the gas capacity factor (gas velocity). When absorber 01 of the schema in Fig. **20b** was sized, the gas capacity factor ($F_V = 1.1$ $Pa^{0.5}$) was higher than 1.0 and very close to the gas capacity factor obtained in simulation results ($F_V = 1.2$ $Pa^{0.5}$). The pressure drop for the real scale column was 145.9 Pa m^{-1}, value close to those expected for absorption columns.

The hydrodynamic behavior of absorber 01 in Fig. **20b** was analyzed with different packing materials. Mc-Pac packing lead to the highest pressure drop and largest diameter (4.5 m), increasing the energy required to move the gases through the column and the amount of packing material required for the same absorber. Envipac packing material would result in lower costs due to the lower diameter and lower pressure drop, as can be observed in Table **12**.

The design of the large-scale absorber 02 (Fig. **20b**) uses the phase ratio determined by the simulation and the gas stream leaving absorber 01. In that case, 30.5 m^3 h^{-1} of water are required to reduce ammonia slip to less than 10 ppm. The flow rates and transport properties for sizing absorber 02 are summarized in Table **11**. Column diameter was 2.0 m with Mc-Pac packing; pressure drop and gas velocity at flooding were 980.5 Pa m^{-1} and 3.7 m s^{-1}, respectively; the gas capacity factor was 3.19 Pa$^{0.5}$. Results for absorber 02 listed in Table **12** do not show major differences in the hydrodynamic behavior of Envipac and Ralu – flow packing: pressure drop, column diameter, liquid hold-up and densities are almost the same. Instead, Mc-Pac packing displays a larger pressure drop and demands a larger diameter. The gas capacity factor of both columns lies within the recommended range of operation and is close to the F_V values obtained by the simulation of the small columns.

CONCLUSION

A non-equilibrium stage, rate-based model for the absorption of CO₂ with ammonia was developed. Simultaneous mass and energy transfer through the gas-liquid interface, as well as the hydrodynamics of packed columns with the corresponding correlations to calculate pressure drop, liquid hold up and mass transfer coefficients, were included in the model which is based on absorption segments. Simulation of the absorption of CO_2, from a mixture with ammonia indicated more than 80% of absorption of CO_2; complete absorption of SO_2 and NO_2 present in the stream could also be accomplished.

It was observed that using a lean solution with low concentration of ammonia would require a large column, increasing packing costs as well as the probability of column flooding. Therefore, a more concentrated aqueous ammonia solution was used, which required implementing an ammonia washer to avoid ammonia slip.

The model was used for sizing a packed column for the treatment of a gas stream with a composition similar to that expected from a wet cement kiln. The best packing options were Envipac or Ralu – flow due to the lower pressure drop and column diameter and their better resistance to corrosive solutions when compared to the Mc-Pac packing material. Besides, the lower density of those materials results in a lighter column. The selected system requires less packing volume, which translates into lower capital costs.

ACKNOWLEDGEMENTS

Financial support of Universidad de Antioquia (Sustainability Strategy 2013-2014) is gratefully acknowledged. A.E. Hoyos thanks Colciencias for a doctoral fellowship and the Laboratory of fluid Sepratations (TU Dortmund) for their hospitality and support during his stay.

CONFLICT OF INTEREST

The authors confirm that this chapter contents have no conflict of interest.

LIST OF SYMBOLS

Symbols	Definitions	Units
a	Specific surface area	$m^2\,m^{-3}$
C_{FL}	Constant for calculating the vapor (gas) velocity at flooding point	dimensionless
C_i	Molar concentration of i	$mol\,m^{-3}$

Cp_G	Heat capacity of vapor	J mol^{-1} K^{-1}
Cp_L	Heat capacity of liquid	J mol^{-1} K^{-1}
C_{Tot}	Total molar concentration in a stream	mol L^{-1}
d_{col}	Diameter of the column	m
D_i	Diffusion coefficient of species i	m^2 s^{-1}
D_{ij}	Binary diffusion coefficient	m^2 s^{-1}
E^L	Rate of energy transfer to liquid phase	kW
E^V	Rate of energy transfer from vapor phase	kW
E_j^V	Molar flow of energy toward the interface	kJ s^{-1}
E_j^P	Rate of energy transfer across the interface (phase P) in stage j	kW
F	Faraday constant	C mol^{-1}
F_V	Gas capacity factor ($U_V * \sqrt{\rho_V}$)	Pa$^{0.5}$
$F_{V,FL}$	Gas capacity factor at flooding ($U_{V,FL} * \sqrt{\rho_V}$)	Pa$^{0.5}$
$f_{i,j}^V$	Flow of species i fed as vapor to the column	kmol s^{-1}
G	Bulk density of packing	kg m^{-3}
G	Molar flow of vapor	kmol s^{-1}
H	Height of the column	m
H_j^V	Enthalpy of the vapor leaving stage j	kJ kmol^{-1}
H_j^{FV}	Enthalpy of vapor fraction in fed stream at stage j	kJ kmol^{-1}
h_L	Liquid hold-up	m
h_{LS}	Liquid hold-up at loading point	m^3 m^{-3}
h^L	Heat transfer coefficient in the liquid phase	kW m^{-2}
h^V	Heat transfer coefficient in the vapor phase	kW m^{-2}
$h_{G\_i}^{in}$	Enthalpy of i in vapor or gas flow entering the stage j	kW

$h_{G\_i}^{out}$	Enthalpy of i in vapor or gas flow leaving the stage j	kW
$h_{L\_i}^{in}$	Enthalpy of i in liquid flow entering the stage j	kW
$h_{L,i}^{out}$	Enthalpy of i in liquid flow leaving the stage j	kW
K1, K2	Exponents for determining the resistance coefficient for single-phase flow of gas in packing for $\text{Re}\,v < 2100$	dimensionless
K3, K4	Numerical value of K1 and K2 for $\text{Re}\,v > 2100$	dimensionless
K_i	Distribution coefficient for the compound i $\left(y_i/x_i\right)$	dimensionless
k_i	Mass transfer coefficient of species i	m s⁻¹
k_{ij}	Mass transfer coefficient for binary mixture of i and j	m² s⁻¹
L	Molar flow of liquid	kmol s⁻¹
L_j	Flow of liquid leaving stage j	kmol s⁻¹
M_S	Molecular weight of the solvent	kg kmol⁻¹
N	Number of packing elements per cubic meter	dimensionless
n	Number of stages	dimensionless
N_i	Flux of component i through the gas-liquid interface	mol s⁻¹ m⁻²
$N_{i,j}^P$	Molar flux of species i through the interface (from phase P) in stage j	kg mol m⁻² s⁻¹
$N_{i,j}^V$	Flow of species i transferred from gas bulk to the interface in stage j	kmol s⁻¹
q^L	Heat flux from/to the liquid phase	kW m⁻²
q^V	Heat flux from/to the vapor phase	kW m⁻²
Q_j^V	Molar flow of energy toward the vapor stream	kJ s⁻¹
R_{ev}	Reynold's number for vapor stream	dimensionless
$R_{i,j}^V$	Production of species i in the vapor phase at stage j	kmol s⁻¹
r_j^V	Fraction of side stream in the vapor phase for the stage j	dimensionless
S	Cross sectional area of the column	m²
T^I	Temperature in the interface	K

T^L	Temperature in the liquid phase	K
T^V	Temperature in the vapor phase	K
U_L	Velocity of the liquid in the column	m s^{-1}
U_V	Velocity of vapor phase in the column	m s^{-1}
$u_{V,FL}$	Velocity of vapor phase in the column at flooding point	m s^{-1}
V_{Bottom}	Actual volumetric flow of gases entering at the bottom	m^3 s^{-1}
V_j	Flow of vapor leaving stage j	kmol s^{-1}
x_i	Mol fraction of i	mol mol^{-1}
Z	Distance in the direction of mass transfer	m
z_i	Electric charge of species i	dimensionless

GREEK SYMBOLS

ΔH_j^V	Production of heat energy due to reactions	kJ s^{-1}
Δp	Pressure drop by meter of packing height	Pa m^{-1}
ε	Void fraction	m^3 m^{-3}
φ_p	Form factor of dry packing	dimensionless
γ_i	Activity coefficient of species i	dimensionless
μ	Dynamic viscosity	mPa s
μ_i	Chemical potential of species i	dimensionless
ρ_V	Density of vapor phase	kg m^{-3}
v_i	Velocity of species i	m s^{-1}
ψ	Resistance coefficient for dry column	dimensionless
ψ_{FL}	Resistance coefficient	dimensionless
ζ_{ij}	Friction coefficient of i with respect to j	s m^{-1}

REFERENCES

[1]   Plaza, J. M., Wagener, D. V., Rochelle, G. T., Modeling CO_2 capture with aqueous monoethanolamine, *Energy Procedia*, 1, pp. 1171-1178, **2009**.

[2]   Kucka, L., Müller, I., Kenig, E. Y., Górak, A., On the modelling and simulation of sour gas absorption by aqueous amine solutions, *Chem. Eng. Sci.*, 58, pp. 3571-3578, **2003**.

[3]   Zhang, Y., Chen, H., Chen, C.-C., Plaza, J. M., Dugas, R. and Rochelle, G. T., Rate-Based Process Modeling Study of CO_2 Capture with Aqueous Monoethanolamine Solution, *Ind. Eng. Chem. Res.*, 48, pp. 9233–9246, **2009**.

[4]   Kooijman, H. A. and Taylor, R., *The ChemSep Book*, 2nd ed.; ChemSep: Delft, pp. 245-258, **2006**.

[5]   Maćkowiak, J., *Fluid Dynamics of Packed Columns: Principles of the Fluid Dynamics Design of Column for Gas/Liquid and Liquid/Liquid Systems*, Springer-Verlag: New York, **2010**.

[6]   Kohl, A. L., In: *Handbook of Separation Process Technology*; Rousseu R.W., Ed.; John Wiley & Sons Inc.: New York, pp. 340-404, **1987**.

[7]   Kolev, N., *Packed Bed Columns for absorption, desoprtion, rectification and direct heat transfer*, ELSEVIER: Amsterdan, **2006**.

[8]   Billet, R., *Packed Towers in Processing and Environmental Technology*, Verlagsgesellschaft mbH: Weinheim, **1995**.

[9]   Raschig-Jaeger-Technologies: Random Packings, [Online] **2012**, http://www.raschig.de/Random-Packings (accesed January 9, 2012).

[10]   J. Environmental: Jaeger Environmental Products, [Online] **2012**, http://www.jaegerenvironmental.com/nor-pac.htm (accesed January 9, 2012).

[11]   E. E. GmbH. McPAc., [Online] **2012**, http://www.envimac.de/58.html?&L=1 (accesed January 9, 2012**).**

[12]   Thiele, R., Faber, R., Repke, J. U., Thielert, H. and Wozny, G, Design of Industrial Reactive Absorption Processes in Sour Gas Treatment Using Rigorous Modelling and Accurate Experimentation, *Chem. Eng. Research & Design*, 85, pp. 74-87, **2007**.

[13]   Green, D. and Perry R., Perry's *Chemical Engineers' Handbook*, 8th ed.; McGraw-Hill Education: New York, **2007**.

[14]   Maćkowiak, J., Determination of Flooding Gas Velocity and Liquid Hold-up at Flooding in Packed Columns for Gas-Liquid Systems, *Chem. Eng. Sci.*, 13, pp. 184-196, **1990**.

[15]   Billet, R. and Schultes, M., Fluid dynamics and mass transfer in the total capacity range of packed columns up to the flood point, *Chem. Eng. Technol.*, 18, pp. 371-379, **1995**.

[16]   Billet, R. and Schultes, M., Modelling of pressure drop in packed columns, *Chem. Eng. Technol.*, 14, pp. 89-95, **1991**.

[17]   Billet, R. and Schultes, M., Prediction of mass transfer columns with dumped and arranged packings: Updated summary of the calculation method of Billet and Schultes, *Trans IChemE*, 77, pp. 498-504, **1999**.

[18]   Wang, G.Q., Yuan, X.G. and Yu, K.T., Review of Mass-Transfer Correlations for Packed Columns, *Ind. Eng. Chem. Res.*, 44, pp. 8715–8729, **2005**.

[19]   Onda, K., Takeuchi, H. and Okumoto, H., Mass Transfer Coefficients Between Gas and Liquid Phases in Packed Columns, *J. of Chem. Eng.* Japan, 1, pp. 56-62, **1968**.

[20]   Aspen Technology Inc., Aspen Custom Modeler: Library Reference Guide; Aspen Technology: Cambridge, **2005**.

[21] Danckwerts, P.V., Gas-Liquid Reactions, McGraw-Hill: New York, **1970**.

[22] Noeres, C., Kenig, E. Y. and Gorak, A., Modelling of Reactive Separation Processes Reactive Absorption and Reactive Distillation, *Chem. Eng. & Processing*, 42, pp. 157-178, **2003**.

[23] Brettschneider, O., Thiele, R., Faber, R., Thielert, H. and Wozny, G., Experimental investigation and simulation of the chemical absorption in a packed column for the system NH_3–CO_2–H_2S–NaOH–H_2O, *Sep. & Purif. Tech.*, 39, pp. 139-159, **2004**.

[24] Prausnitz, J.M., Lichtenthaler, R.N. and Gómez de Acevedo, E., *Termodinamica Molecular de los Equilibrios de Fase*, Prentice Hall Iberia: Madrid, **2000**.

[25] Chen, C.-C., Britt, H. I., Boston, J. F. and Evans, L.B., A Local Composition Model for Excess Gibbs Energy of Electrolyte Systems Part I: Single Solvent, Single Completely Dissociated Electrolyte Systems, *AIChE J.*, 28, pp. 588-596, **1982**.

[26] Chen, C.-C. and Evans, L.B., A Local Composition Model for the Excess Gibbs Energy of Aqueous Electrolyte Systems, *AIChE J.*, 32, pp. 444-454, **1986**.

[27] Mock, B., Evans, L.B. and Chen, C.-C., *In: Phase Equilibria in Multiple-Solvent Electrolyte Systems: A New Thermodynamic Model, Proceedings of the 1984 Summer Computer Simulation Conference*, Boston July 23-25, 1984; Society for Computer Simulation: La Jolla, Calif. (USA), p 558, **1984**.

[28] Mock, B., Evans, L.B. and Chen, C.-C., Thermodynamic Representation of Phase Equilibria of Mixed-Solvent Electrolyte Systems, AIChE J., 32, pp. 1655-1664, **1986**.

[29] Soave, G., Equilibrium Constants from a Modified Redkh-Kwong Equation of State, *Chem. Eng. Sci.*, 27, pp. 1197-1203, **1972**.

[30] Redlich, O. and Kwong, J. N. S., On the Thermodynamics of Solutions. V An Equation of State. Fugacities of Gaseous Solutions, *Chem. Rev.*, 44, pp. 233-244, **1949**.

CHAPTER 8

Analysis of CO_2 by Determination of Carbon and Oxygen Using Ion Beam Analysis (IBA)

Juan A. Aspiazu[1,*] and Arturo Aspiazu[2]

[1]*Instituto Nacional de Investigaciones Nucleares Carretera Mexico-Toluca S/N, La Marquesa, Ocoyoacac, 52750, Mexico and* [2]*Faculty of Chemistry, National Autonomous University of Mexico, Av. Insurgentes Sur 4411 Ed 25-301 Col. Tlalcoligia, 14430, Tlalpan D.F., Mexico*

Abstract: In several studies related to atmospheric emission of anthropogenic CO_2, it has been established that the resulting greenhouse effect is a direct factor in the climate change observed around the world. These climate changes are a consequence of the influence of the greenhouse effect on the atmospheric thermodynamic state. The importance has been recognized of measuring the atmospheric CO_2 contamination with precision. Also, for a proper disposal of this contaminant, there have been developed some efficient physicochemical procedures for either CO_2 dissociation or recycling. Ion accelerators provide a suite of techniques, collectively referred to as IBA, offering excellent options for the analysis of this kind of contamination.

Keywords: Carbon dioxide, analysis by Ion Beam, analysis for developments on capture and storage technologies.

1. INTRODUCTION

In this chapter, IBA techniques, associated with ion accelerators, will be introduced as useful tools to determine the content of Carbon (C) and Oxygen (O), presumably associated with an atmospheric pollutant like CO_2, when released into the atmosphere by industrial activity. In particular, it is important to determine the Carbon and Oxygen content due to CO_2, when it is an industrial waste product obtained during CO_2 capture by means of a physicochemical process for its permanent storage [1]. Later on, it will be explained how one can characterize those CO_2 capture processes using IBA techniques. In that way, one will be able

*Corresponding author **Juan A. Aspiazu:** National Institute of Nuclear Research, Accelerators Department, Nuclear Center "Dr. Nabor Carrillo Flores", México-Toluca Road, La Marquesa, Zip Code 52750, Ocoyoacac Estado de Mexico, Mexico; Tel: +52 5553-297200; E-mail: juan.aspiazu@inin.gob.mx

Rosa-Hilda Chavez and Javier de J. Guadarrama (Eds)

to make a qualified opinion about the efficiency and efficacy of the selected physicochemical process.

A proper management preventing significant CO_2 pollution will allow the implementation of an environmental policy oriented to avoid the undesirable climate change. This becomes more urgent after realizing that, as is the case for several anthropogenic gases, CO_2 accumulates in the upper atmospheric layers. In this condition, this atmospheric layer absorbs infrared radiation from the Earth, reflecting (*or reemitting*) it back to the ground. The perturbation due to the additional absorbed energy is enough to rise the atmospheric temperature, occasionally producing atmospheric turbulences, that is one cause of Earth's climate changes.

There are many laboratories with ion accelerators capable of providing ions with energies of several tens of MeV. These ions beams can be used for the elemental analysis of different kind of samples of interest in basic or applied research. In this way, the entire periodic table can be covered when analyzing a particular sample. Also, wellknown interaction processes taking place between the energetic ions and matter allow us to identify the sample atoms, as well as to determine their concentrations. All one needs to know for this kind of analysis is the interaction probability (*cross section*) between the *"projectile"* ion and the sample *"target"* atom. This probability depends on the projectile energy and the atomic number of the target atom.

2. BASIC ION-MATTER INTERACTIONS

The ion-matter interaction can manifest itself on the projectile as: energy degradation, angular dispersion, capture or nuclear reaction with the target. In this last case the projectile may undergo a nuclear transmutation, *i.e.*, may change its isotopic composition. Another possible result of the interaction may be an *"elastic collision"*, in which the relative speed between the target and projectile is conserved, but not necessarily its direction, which provides important information about the target atom. Also, it is possible that, after the reaction takes place, some part of the relative kinetic energy between the two colliding particles shows up as internal energy (*excitation energy*) of the target atom. As a consequence of this energy absorption by the target it gets to an excited state that normally last a very tiny time, eventually decaying through an atomic or nuclear process into a more stable state, for example, the ground state of the atom or nucleus. As a consequence of this atomic or nuclear de-excitation process, a particle or a photon may be emitted as secondary radiation, which is characteristic of that process. All these possible processes follow quite precise statistical laws with very well defined probabilities of occurrence for each one.

According to this, in principle, measuring the energy spectrum of secondary radiation, one can evidence the presence of a particular element in the analyzed sample. Additionally, determining the number of events under a selected peak (*peak area*) appearing in the measured energy spectrum (*selecting the energy*), and knowing the number of projectiles that traversed the sample during a predefined time interval (*charge integration*), and using the known relevant reaction cross section values, one can calculate the corresponding number of target atoms in the sample, *i.e.*, its concentration [2].

3. IBA TECHNIQUES

Quantitative analysis requires a *"homogeneous"* sample, so one has to keep this in mind when preparing the target sample prior the experiment. Using IBA techniques one can determine nominal concentrations with a precision of parts per million (ppm), but using target concentration, it is possible to get a precision down to parts per billion (ppb) or, in some cases, two orders of magnitude lower. The normal uncertainty in concentration measurements is on the order of 5%. Fig. **1** shows a typical IBA experimental setup.

IBA techniques allow simultaneous elemental analysis, which complement themselves covering all elements. They have the advantage of being more accurate than conventional chemical analytical techniques: Titrimetry, Electro-analysis, Spectroscopy and Chromatography. When using IBA, a normal data collection time is about 10 minutes for a homogeneous and solid sample. In case of liquid or gaseous samples, the analysis is relatively more difficult due to a more complex sample preparation, and also due to a more complex experimental setup inside the target vacuum chamber. When dealing with solid samples, one can obtain precise elemental concentration surface profiles down to a monolayer resolution. One can get depth resolution up to a few nanometers and atomic resolution using specialized equipment. Using an IBA technique called channeling to study damaged mono-crystalline structures it is possible to determine the spatial profile of the damages.

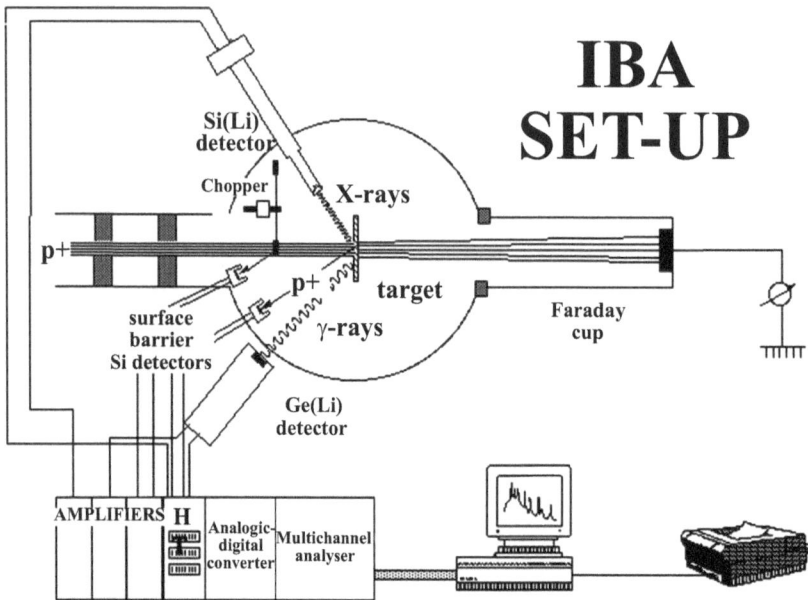

Figure 1: IBA experimental set up

Each IBA technique is associated to a specific kind of projectile, its energy, as well as the corresponding secondary radiation resulting from the projectile-target interaction. As a result of this, each IBA facility is specialized on certain kinds of applications, mainly when one considers the detection capability (*sensitivity*) required by the kind of studies corresponding to a specific research field (Fig. **2**). To clarify this, a brief summary of some IBA techniques together with their applicability areas is presented below:

- *"Rutherford Backscattering" (RBS): Suitable for analyzing the "heavy" element content in a "light" matrix containing elements with atomic number Z ≤ 8. This technique uses a simple assumption that models the target-projectile interaction as a "hard-sphere" elastic collision.*

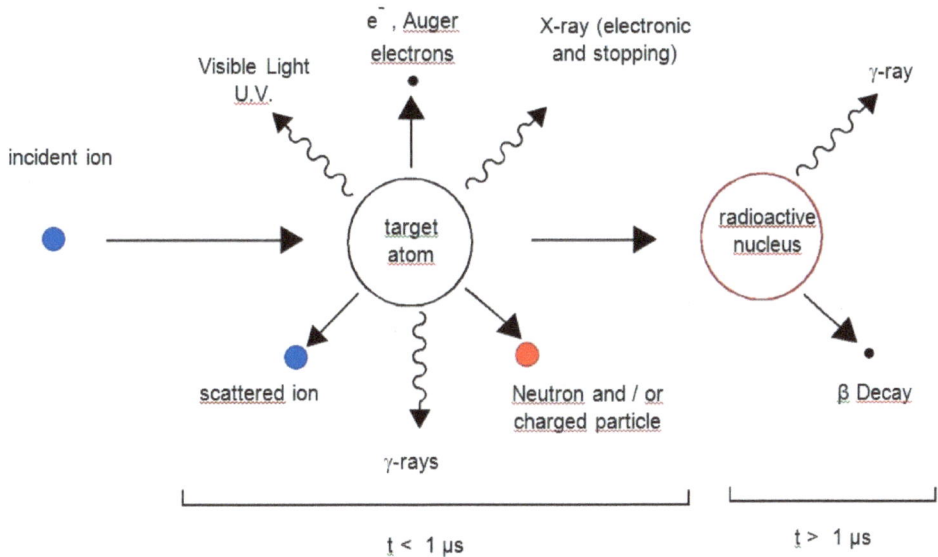

Figure 2: Secondary radiation induced by energetic ions interaction.

- *"Elastic (non-Rutherford) Backscattering Spectrometry" (EBS):* Suitable for determining light elements implanted in a heavy element matrix. In this case, it is assumed that the projectile energy exceeds the "Coulomb barrier" of the target atomic nucleus. For this reason, it is necessary to use the Schrödinger wave equation to obtain the scattering cross section.

- *"Elastic Recoil Detection Analysis" (ERDA):* Suitable for light element determination in a heavy matrix. This technique is based on describing the collision as an "elastic" nuclear interaction.

- *"Particle Induced X-Ray Emission" (PIXE):* Suitable for the analysis of elements with $Z \geq 12$ present in a sample as components or as "traces". In this kind of interactions, the projectile transfers a certain amount of energy to an orbital electron of the target nucleus that is larger or equal to the electron's binding energy. In this way, the electron will be ejected leaving behind a "vacancy". This vacancy in the target atom will be occupied by means of an electronic transition from an "outermost orbital" which will be accompanied by *the emission of a characteristic X-ray with an energy value equal to the difference of the two electron orbital energies.*

- *"Nuclear Reaction Analysis" (NRA): Suitable for isotopic characterization of samples. This technique is based on selecting a specific reaction, depending on the target isotope that needs to be identified, that presents a strong resonance (strong enhancement of the cross section value) at a well-defined "resonant energy". At this resonant energy, the target nucleus will be promoted very favorably to an unstable excited state. Eventually, the excited nucleus will decay to its ground state, simultaneously emitting a characteristic radiation (usually ionizing radiation).*

- *"Channelling": Suitable for detection of "damage" in a crystal lattice. When the path of an energetic ion incident upon the surface of a monocrystal lies close to a major crystal direction, the particle will, with high probability, suffer small-angle scattering as it passes through several layers of atoms in the crystal. So if, instead, one observes strong backscattering of the bombarding ion, it will be a signature of the presence of "damage" in the crystal lattice, i.e., interstitial atoms, "dislocations" (line defects).*

There exist other possible types of analysis based on additional techniques like *"Particle Induced Gamma Ray Emission"* (PIGE), *"Deuteron Induced X-Ray Emission"* (DIXE), *"Heavy Ion Backscattering"* (HIBS), *"Time of Flight"* (TOF), *etc.*, increasing the suite of available IBA techniques.

4. ION PRODUCTION

The *"ion beams"* used in IBA techniques are made in several steps. In the first step, one uses a radio-frequency or a sputtering ions source to obtain ions. For example, in the latter case one uses a cone made out of the specific material (*chemical element*) from which one needs to get the ions. In the second step, since this ion production is not a clean process (*different kinds of ions are produced*), an electromagnet (*injection magnet*) is used to get rid of all undesirable ion species.

In the third step, once the ion beam is *"clean"*, it is *"injected"* into the accelerator tank, which is pressurized up to 14 atm. In the accelerator tank, the ion beam travels inside a device called an accelerator tube, which is under high vacuum ($\sim 10^{-6} Torr$). These tubes are made of a set of accelerating plates or electrodes that establish an electric field ($\sim 1 MV/m$) responsible for accelerating the ions up the desired energy (*a few MeV/nucleon*).

Since the *"ion beams"* obtained in this way are made of diverging ions, it is necessary to counter the divergence with a special focusing magnetic devices called electromagnetic quadrupole lenses. Their function is similar to what an optical lens does when it is used to focus a light beam. During the *"optic"* design of the *"beamline"*, one determines the optimal locations for the electromagnetic quadrupole lenses along the beamline.

At this point one has an energetic ion beam but its energy is not quite well defined. For that reason a special electromagnet (*analyzer magnet*) is used, the purpose of which is to select ions with a velocity that falls inside of a very small velocity window, which corresponds to a very small energy window or energy resolution. At this point the beam is ready to *"bombard"* the sample inside the irradiation chamber.

5. SECONDARY RADIATION DETECTION AND SIGNAL ANALYSIS

As a result of the target-projectile interaction, secondary radiation is emitted from the analyzed sample. This secondary radiation will deposit all or part of its energy in a *"detector"*. Since the nature of this secondary radiation is diverse, then one needs to use the right detector. For electromagnetic radiation like X-rays or Gamma rays one can use detectors of doped silicon with lithium*Si(Li)*, doped germanium with lithium *Ge(Li)*, hyper pure germanium *HPGe* or *"scintillators"*. For charge particles one can use *"surface barrier"* detectors.

When secondary radiation impinges in one of these detectors a signal is generated by the collection of a small amount of charges created inside the detector. This signal, before being usable, has to be carefully processed by specialized electronics; a preamplifier and an amplifier are required for this purpose. In this way one gets a voltage pulse (*electronic signal*) whose amplitude is determined by the amount of energy deposited at the detector by the secondary radiation. Then, these pulses are processed by a data acquisition device (DAQ), and, finally, it saves the processed data for its eventual off-line analysis. Except for the preamplifier, all the electronics are normally in a control room, out of the bombarding room.

The way in which the DAQ works is by building, for each relevant signal or parameter, a histogram called its spectrum. For example, an energy spectrum is a two dimensional histogram obtained after digitizing the pulse height of each energy event and saving it in a channel that falls in a specific small energy interval represented by that channel. In this way, the spectra obtained for all relevant measured parameters of all secondary radiations emitted during the *"bombarding"* process constitute an *"atomic-nuclear profile"* of the studied sample.

The next and last step is the interpretation of all the spectra, which depends on the particular selected IBA technique. This means that the form of an energy spectrum will depend on the selected IBA technique. For example, in the case of a PIXE spectrum, it will consist of all the peaks or lines corresponding to all the characteristic X-rays emitted by elements in the sample with atomic number (Z) > 12. There is a double task during the analysis. One has to perform element identification and also calculate its concentration. It may occur that two peaks are overlapped. In that case one has to *"deconvolute"* those peaks in order to be able to correctly determine the *"area"* of each peak (*total number of counts under the peak*), as well as their energies. Beside this, the number of counts in a peak ought to be large enough in order to warrant a small uncertainty, ensuring that the peak stands out from the background.

Depending on the nature of the analyzed sample, especially in relation to its homogeneity, well known formulas can be used to obtain the concentrations of the different elements present in the sample. Nowadays, there are many commercial analysis software packages for nearly all IBA techniques. The software package "GUPIX" is adequate software for PIXE analysis. However, in many cases, due to the complex analysis required, it is the researcher who has to develop the codes to analyze the spectra.

6. ANALYSIS CARBON-OXYGEN

Three techniques, RBS, NRA and ERDA, are commonly used for carbon analysis due to their analytical capabilities which have significant advantages in most cases. RBS has a detection capability> 10^{-4} for atomic fraction, surface analysis < 10^3 nm resolution in depth < 10^2 nm and resolution for atomic number of 1; NRA has a detection capability > 10^{-3} for atomic fraction, surface analysis <10^4 nm and depth resolution < 10^2 nm; and ERDA has detection capability > 10^{-4} for atomic fraction, surface analysis < 10^2 nm and depth resolution < 10 nm. For oxygen, RBS and ERDA are used with same advantages that were indicated for carbon.

The following Figs. **3** and **4** show some examples for carbon and oxygen. Given the nucleus to be analyzed and a specific projectile, the graphs indicate the value of the interaction section as a function of projectile energy (*"excitation curves"*), as well as the nucleus and particle resulting from the interaction and, the detection angle considered for such determinations.

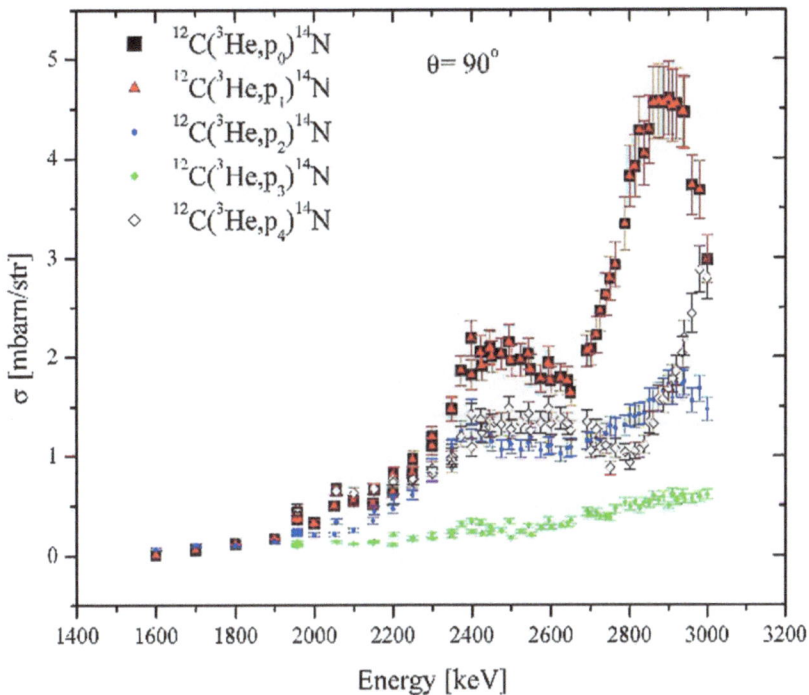

Figure 3: Example of determining ^{12}C by producing protons (p$_i$) with defined energies, and ^{14}N when it is irradiating at different energies of ^{3}He at fixed angle 90 °. [3].

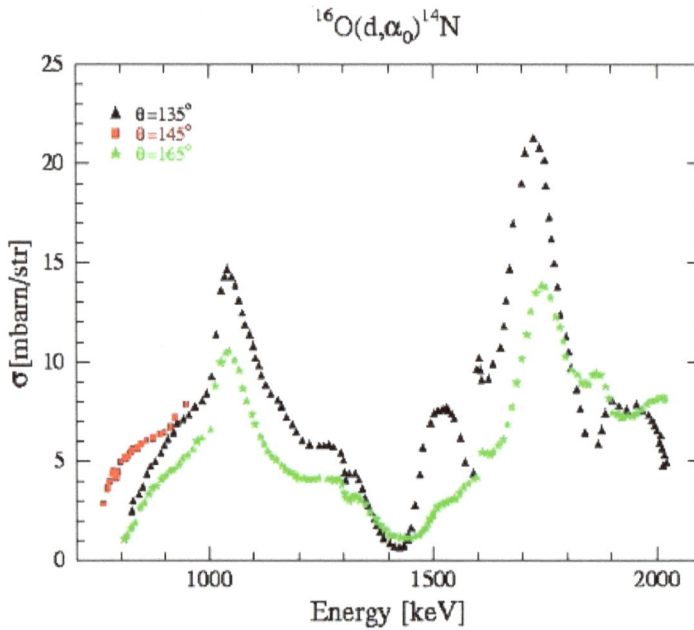

Figure 4:. Example ^{16}O determination by producing alphas (α_0) and ^{14}N when it is irradiate at different deuterium (d) energies and different angles.

Using this information, one can then plan the experiment accordingly by selecting the experimental conditions that favor obtaining useful results for the analysis that is being performed. However, there are cases where the analysis is restricted by specific conditions on these variables, and thus the time to perform a proper analysis needs to be estimated as an essential part for planning the experiment. Knowledge of the interaction section and maximum intensity of *"beam"* projectiles (*"beam current"*) that the system can achieve, allow estimation of the minimum time to complete analysis.

7. TECHNOLOGIES FOR MANAGEMENT OF CO₂

This section refers, in a general way, to technologies that are now most commonly used in capture, storage and recycling of CO₂ [4]. The main purpose is to indicate the nature of samples that would eventually have to be analyzed by IBA techniques that, due to characteristics described in previous sections, can provide a fundamental basis for determination of physical and chemical phenomena that take place during realization of processes involved, with the consequence that it is possible to obtain higher accuracy to estimate the efficiency of such processes. All types of laboratory studies open up new possibilities for creating innovative technologies in the field, especially with regard to CO_2 in searching for processes with better energy-efficiency, at minimum costs, and greater environmental protection too. Particularly, in relation to preventing environmental damage, one must reject processes which have, as waste products, material contaminating for the atmosphere [5, 6].

7.1. CO_2 Capture Processes

Global CO_2 emissions currently account for more than 30,000 million metric tons/year, but the capabilities developed so far for capture and storage reach only 5 million metric tons/year (Fig. **5**).

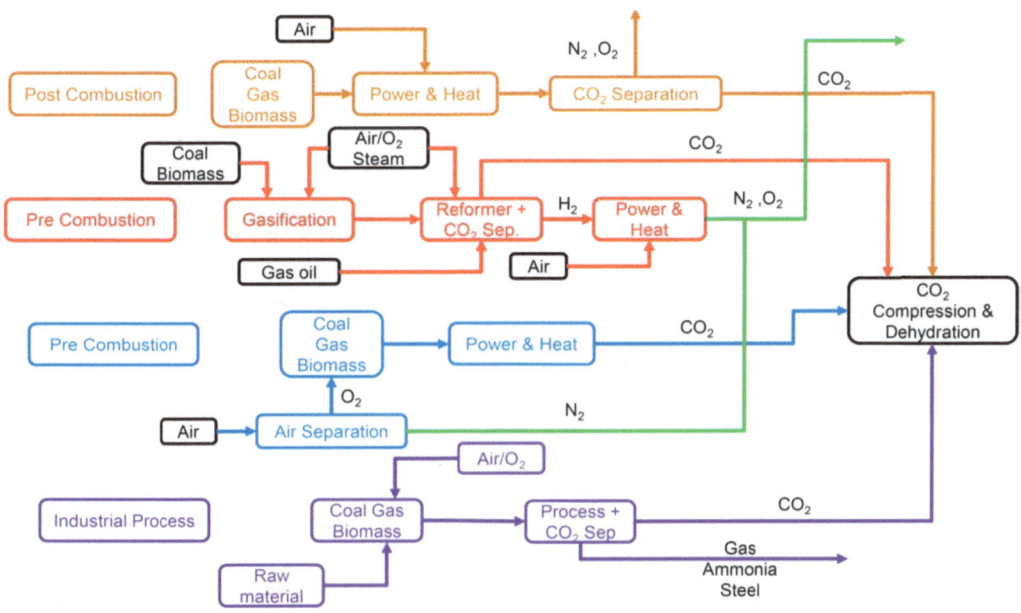

Figure 5: Overview of CO_2 capture processes and systems.

This 0.02% does not permit us to think, for the short term, on a balance in this regard. So, hence the urgency to increase installed CO_2 capture and storage capacity and to develop new technologies with greater advantages [7, 8].

Technologies for capture and storage of CO_2 (Fig. **6**), mainly from power plants producing electricity, are seen as critical to stop risingglobal temperatures [9]. For the more than 20,000TWh of electricity production worldwide, about 42% is obtained using coal, 21% using natural gas and 6% using petroleum, while the rest (31%) comes from nuclear, hydroelectric and renewable sources [10, 11]. Balance will also require the capture and storage of CO$_2$ emissions from other centralized sources dedicated to fuel processing and others sources located in the industrial sector [12]. The capture of CO$_2$ from centralized sources, is followed by the transport of CO$_2$ using pipes, vessels or appropriate vehicles, to sinks used for storage through injection of CO$_2$ into underground geological formations, such as depleted deposits of oil or gas, storage in oceans, *etc.* [13, 14].

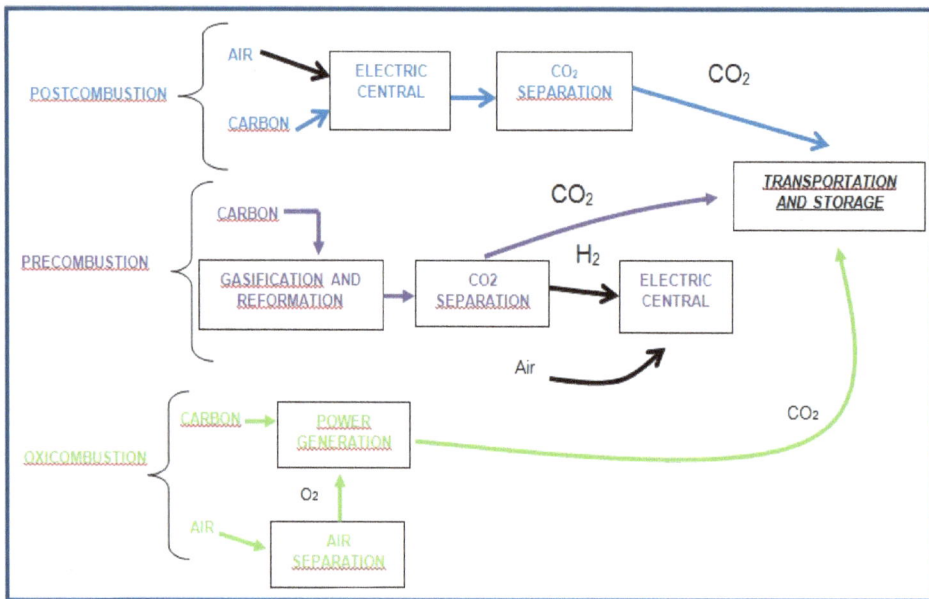

Figure 6: Settings capture systems.

The capture of CO$_2$ involves its separation (Fig. **7**) from flue gases where it is generally present in an amount less than 20% [15-18]. At present, the technologies that are mainly used for CO$_2$ capture are based on:

-   **Post-Combustion:** Cycles of absorption/desorption operating chemically in one direction and in reverse sense to give a gas of high CO$_2$ content (Fig. **8**).

The main chemical equation which governs the absorption process is:

$$H_2O + CO_2 + MEA \rightarrow MEA\,CO_2- + H_3O^+$$

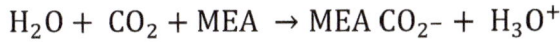

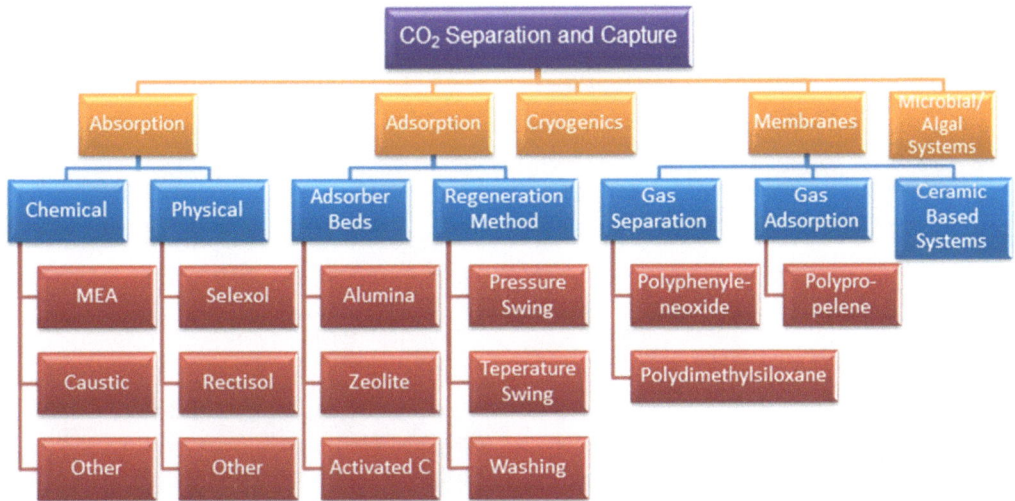

Figure 7: Technical Options for CO_2 Capture.

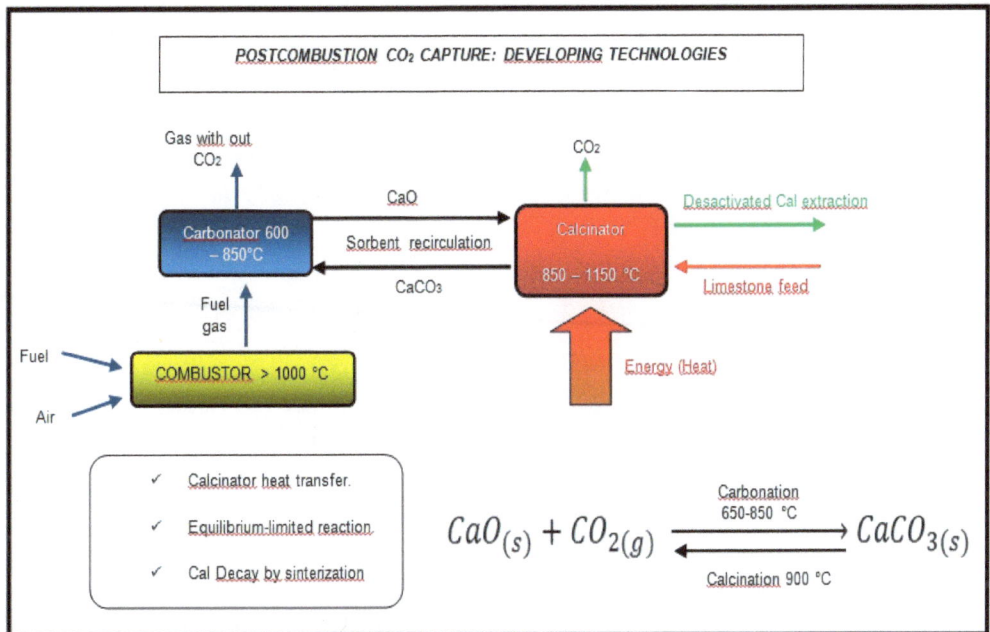

Figure 8: Post-Combustion CO_2 capture: Developing technologies.

where MEA (*mono-ethanolamine*) is an *"amine"* which functions as alkali base in an aqueous medium containing an acid gas (Fig. **9**). The main chemical equations for the solvent regeneration are:

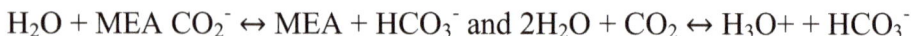

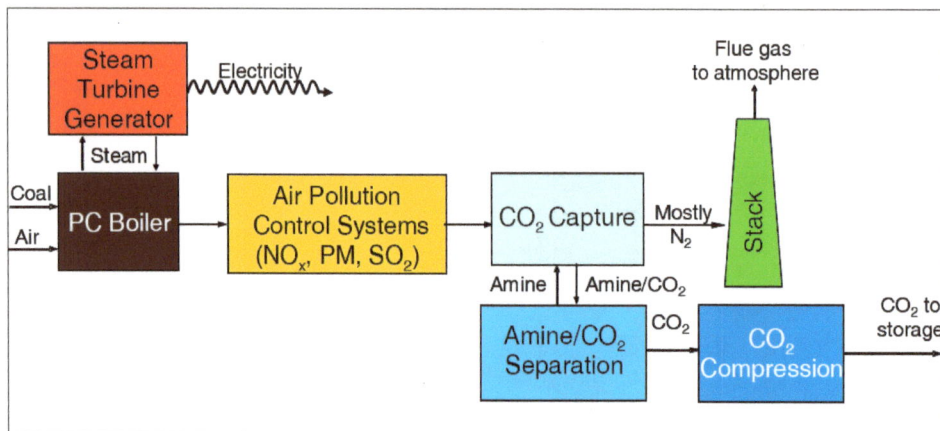

$$H_2O + MEA\ CO_2^- \leftrightarrow MEA + HCO_3^-\ \text{and}\ 2H_2O + CO_2 \leftrightarrow H_3O^+ + HCO_3^-$$

Source: E. S. Rubin, "CO₂ Capture and Transport," *Elements*, vol. 4, (2008), pp. 311-317.

Figure 9: Schematic of a Coal-Fired Power Plant with Post-Combustion CO₂ Capture Using an Amine Scrubber System.

-    *Pressure Adsorption "Swing" (PSA) is the most widely used method to purify synthesis gas to obtain pure H₂. However, the method is not selective when separating CO₂ from remaining purged gases, so that an additional PSA unit prior to H₂ separation would be required, to separate the CO₂ [19].*

-    *Chemical Processes with Adsorbents, where chemical solvents are also used to remove CO₂ from the synthesis gas at partial pressures below 1.5 megapascal (MPa) and are similar to other methods used in post-combustion capture. The CO₂ is removed from the synthesis gas, after a "shift" process:*

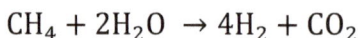

$$CH_4 + 2H_2O \rightarrow 4H_2 + CO_2$$

by means of a chemical reaction that can be reversed by reduction of pressure and temperature [20-22].

- **Physical Processes with Adsorbents,** *mainly apply to streams of gases with high total pressures or high partial pressures of CO$_2$. The solvent regeneration is produced by release of pressure in one or more stages.*

- **Improved Adsorption Reaction (SER),** *is the use of a "packed bed" containing a mixture of catalyst and selective adsorbent to remove CO$_2$ from the reaction zone at high temperature. The adsorbent is periodically regenerated through a temperature (steam) "swing" adsorption system, or by pressure.*

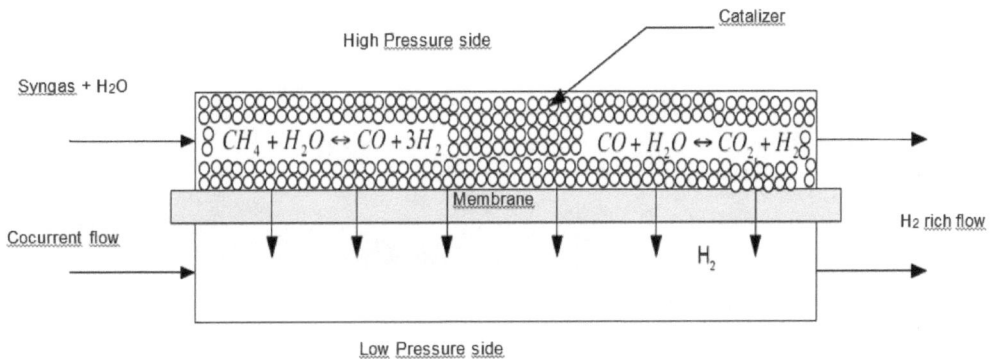

Figure 10: Operation principle of membrane reactor.

- **Membrane Reactors for Hydrogen Production with CO$_2$ Capture.** *Some inorganic membranes offer the possibility of combining processes of reaction and separation, at high temperature and pressure (Fig. 10).*

There are several types of inorganic membranes that permit the separation of H$_2$O$_2$ (Fig. **11**) and CO$_2$. The membranes are divided into two categories: dense membranes and porous membranes, according to their structural characteristics, which may have a significant impact on their performance as separators and/or reactors.

Among the inorganic membranes that allow separation of gases are found micro-porous membranes for separation of hydrogen or carbon dioxide, dense metallic membranes for hydrogen separation and dense membranes based on mixed conductors for hydrogen separation.

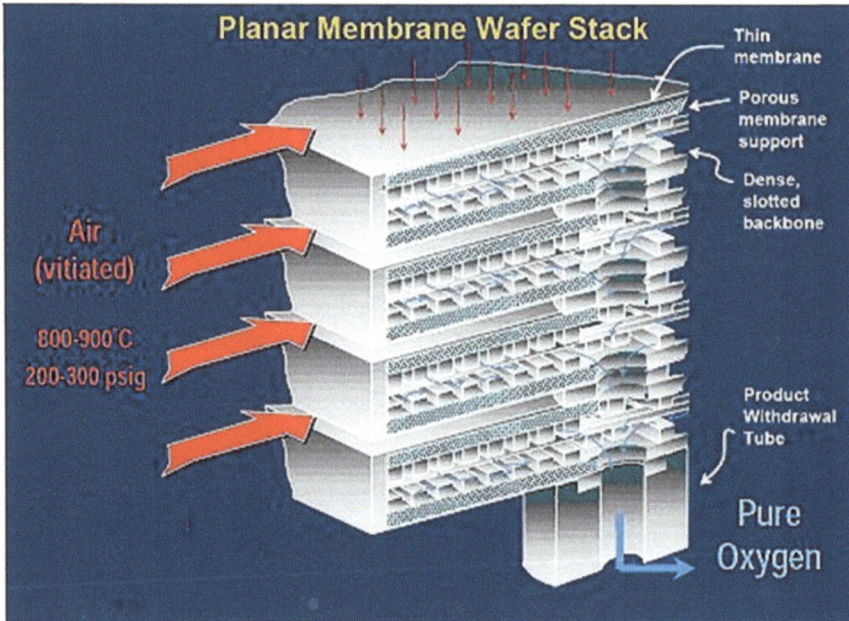

Figure 11: The Ion Transport Membrane (ITM) Oxygen Production Technology Being Developed by Air Products.

Particularly, the micro-porous membranes used for separation of hydrogen or carbon dioxide are zeolite based membranes, which are poly-crystalline thin films supported on rigid porous substrates. Examples of these are macro-porous ceramics and meso-porous or stainless steel.

The amorphous glass (*silica*) contains mostly irregular pores defined by six tetrahedrons SiO$_2$ connected, with an average pore size of about 3Å. These membranes are perm-selective for small molecules perm-selective, such as hydrogen and helium.

The main technical challenge in this area is to adapt *"After-burning Technology"* to a large coal power plant which already exists [23, 24]. In addition, a great effort is focusing on developing more effective catalysts that are stable to pollutants in gases from emissions, as well as to high temperatures, and that are also of lower cost compare to those based on MEA (*mono-ethanolamine*) [25]. This effort is applied not only to the catalysts, but also to chemicals for dissolution and absorption of CO_2, membranes, materials that react with CO_2 and simultaneously fix it permanently, *etc.*

- **Pre-Combustion:** CO_2 separation from gasifiers, before insertion of synthesis gas to gas turbine. These are processes of CO_2 capture prior to combustion, in a gas stream whose main components are CO_2 and H_2, where those can easily be separated. *"Pre-combustion"* capture technologies can be applied to fossil resources such as natural gas, fuel oil and coal, as well as to biomass and wastes [26, 27].One can distinguish three main steps in harnessing of primary fuels with pre-combustion capture:

a) *Reaction of synthesis gas production (These processes lead to generation of a stream composed primarily of hydrogen and carbon monoxide, from primary fuel);*

b) *Reaction "shift" to convert the synthesis gas CO to CO_2 (This reaction provides more hydrogen to the gas stream of the previous phase);*

c) *Separation of CO_2 (There are various methods to separate CO_2 from a stream of CO_2 and H_2). The CO_2 concentration in the input stream to a separator can be between 15-60% dry basis and stream pressure of 2-7 MPa. The separated CO_2 is available for storage.*

Some processesused:

- *Coal Gasification, Petroleum Residues and Biomass,* is basically partial oxidation of fuels, with the peculiarity that in many cases it also supplies steam to the reactor.

- *Technologies Based on Calcium Oxides,* are pre-combustion systems based on the carbonation reaction of calcium oxide ("lime") at high pressures and temperatures. This addition integrates the fuel gasification, conversion reaction "shift" and elimination "in-situ" of CO_2 with CaO. The overall reaction to be achieved is the *regeneration of "sorbent" by calcination of limestone which produces high purity CO_2.*

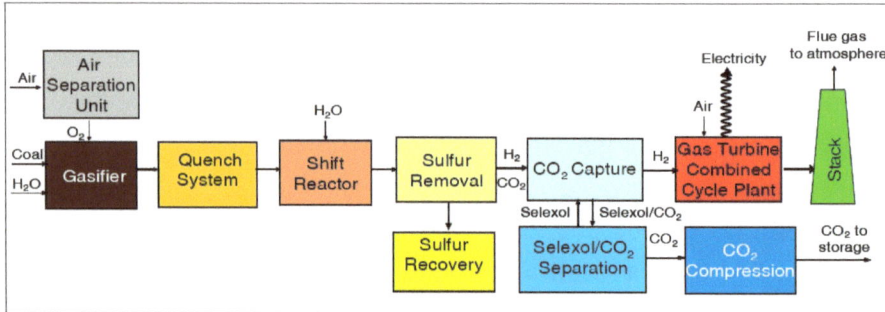

Figure 12: Schematic of an integrated Gasification Combined Cycle (IGCC) Coal Power Plant with Pre-Combustion CO_2 Capture Using a Water-Gas Shift Reactor and a Selexol CO_2 Separation System.

The research on pre-combustion capture processes seeks mainly to reduce the large amount of additional power needed to invest in those processes. Therefore, the main effort is focused on researching new routes on thermodynamic phase diagrams, corresponding to capture and subsequent release of CO_2. One effort also focuses on development of ultra-thin membranes of high flux and maximum selectivity (Fig. **12**).

1) **Oxy-Combustion:** Oxidizer combustion with high oxygen content and a very low presence of nitrogen so that concentration of CO_2 in gases is very high. Before reference to various advanced technologies in the scope of oxy-combustion existing to debug gas streams, the first question is to know the origin of non-CO_2 compounds, like:

a) *Fuel derivatives: H_2O, CO, SO_x, NO_x, H_2S, HCl, HF, H_2, CH_4, heavy metals, hydrocarbons, particulate matter;*

b) *Derived from used air or other oxidant for fuel combustion: O_2, N_2, Ar;*

c) *Derivatives of air leaks in CO_2 capture system CO_2 when operating at sub-atmospheric conditions: O_2, N_2, Ar; or*

d) *Derivatives from CO_2 capture processes or cleaning CO_2: NH_3, solvents.The components and concentrations are dependent on the capture process used, and also on cleaning processes selected for CO_2.*

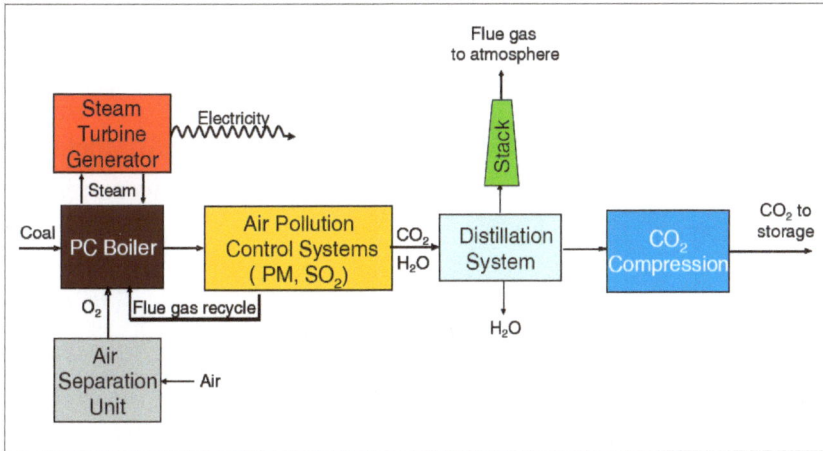

Source: Rubin, "CO₂ Capture."

Figure 13: Schematic of a Coal-Fired Power Plant Using Oxy-Combustion.

Oxy-combustion technology is used in industries such as aluminum, glass and steel for commercial implementation of CO_2 capture technologies and processes (Fig. **13**), but development is still insufficient [28].

- ***Application of Oxy-combustion to Centers of Electricity Production of Type Critical and Supercritical:*** *In these processes, the oxy-fuel combustion chamber provides heat through a heat exchanger to another fluid. Hydrocarbons such as coal can be used in these applications* [29].

Since being proposed in 1982, and its development encouraged by promising technologies created during that time for sequestration of CO_2 in power plants using pulverized coal, oxy-combustion has attracted great interest in studies worldwide [30-32]. Laboratory scale studies cover many fundamental scientific and engineering aspects in application of this technology, mainly on characteristics for coal reactivity and combustion, heat transfer and emissions.

Oxy-combustion is the option that is considered most promising for application in industry, especially in the cement industry, because it can reduce CO_2 capture costs to less than half in relation to the well-known process of after-burning. Also under study are questions of how to reduce excess oxygen required in the process and of identifying uses for surplus nitrogen. The most profitable strategy for this technology will require the development of advanced materials that do not deteriorate at high temperatures in oxygen-rich atmospheres.

2) **Loop Chemical (*"Chemical Looping"*):** The use of metal oxide particles as oxygen carriers which perform a cycle between reactors of combustion and air [33].

- *Gasification by "Chemical Looping" is another system that is investigated for the production of synthesis gas. When the amount of oxygen supplied by the metal oxide to the reactor reducer is lower than the stoichiometric, fuel reaction occurs to CO and H2, and CO is converted to CO$_2$ [34].*

- *Combustion "Chemical Looping" is a process based on oxidation and reduction of metals such as iron, nickel, copper or manganese (Fig. 14). The process involves passing a stream of air through a reactor containing the powdered metal (100 - 500 microns) at a temperature of between 800 to 1,200 ° C. The metal is oxidized and passes into another reactor where it reacts with fuel, is reduced and also provides the necessary oxygen to perform combustion. This technology is at laboratory level with pilot plants of a few kW. It uses alloys of NiO, Co$_3$O$_4$/CoAlO$_4$, NiO/MgAl$_2$O$_4$, Mn$_3$O$_4$/Mg-ZrO$_2$, Fe$_2$O$_3$/Al$_2$O$_3$, and CuO. The hours of operation of these pilot plants are around 1,000 h. [35].*

This technology is being addressed to processes that are applied on mineral carbonation, so that the reaction product itself generates solid material with added value, such as high surface area silica, iron oxides and magnesium carbonate, while safe and durable storage of CO_2 is achieved. Technology also seeks that use of nano-iron oxide particles may serve to convert synthesis gas into high purity hydrogen while CO_2 is captured. The process is considered especially for synthesis gas produced from biomass.

Finally, many simulation studies are being conducted in order to study the dependence of the efficiency of the process with respect to reaction temperature, gas flow and particle size [36].

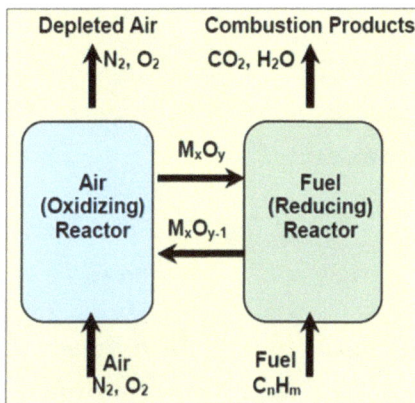

Figure 14: Schematic of a Chemical Looping Combustion System.

7.2. CO₂ Storage Processes

A captured gas with high content of CO_2 is compressed to at least 200 bar, and then injected into suitable storage, for which the following options are considered:

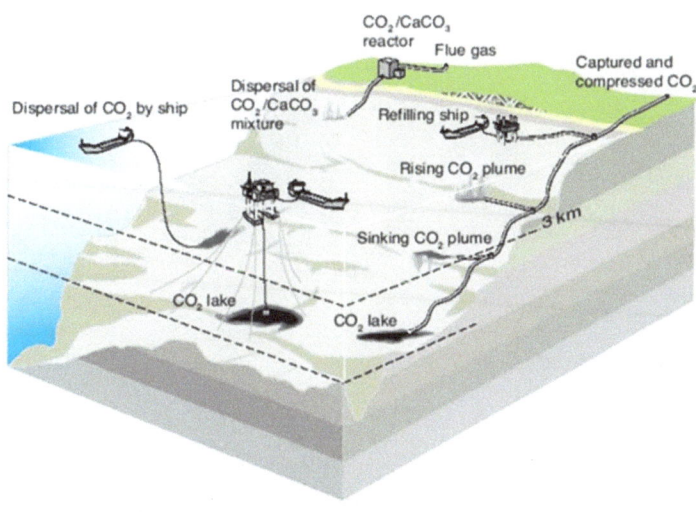

Figure 15: Methods of ocean storage.

*i). **Geological Structures:*** *That contained oil or gas or other similar geological structures.*

*ii). **Saline Aquifers:*** *Deep geological structures with porosity and features that enable the presence of saline water, where CO_2 can be injected so that it would be dissolved in the aquifer permanently.*

*iii). **The Oceans:*** *Artificial addition of CO_2 into the oceans may have huge environmental impact but nonetheless is currently being considered in a rigorous scientific manner (Fig. **15**).*

*iv). **Cavities Created by Dissolving Salt:*** *For this type of storage, the cavity could be generated artificially in a salt formation.*

*v). **Afforestation and Agro-Energy Techniques:*** *Given the important role they can play as CO_2 sinks, they are also considered as an option [37, 38].*

7.3. Research of New Lines for Developments and Prospects for CO_2 Capture and Storage (CCS) (Fig. 16)

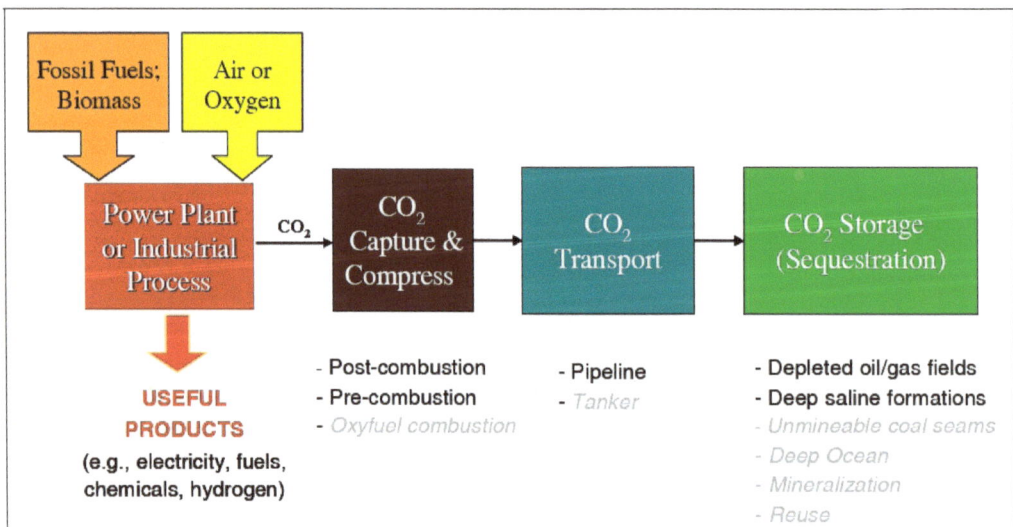

Source: E. S. Rubin, "Will Carbon Capture and Storage be Available in Time?" Proc. *AAAS Annual Meeting, San Diego, CA, 18-22 February 2010*, American Academy for the Advancement of Science, Washington, DC.

Figure 16: Schematic of a CCS System, Consisting Capture, Transport and Storage of CO_2.

CCS Through Nanotubes: Gas separation membranes based on nanotubes are developed on an industrial scale, and it is expected that first prototypes can be applied soon to separation of CO_2 [39]. These membranes permit selective diffusion of CO_2 much more freely (*at speeds greater than 100 times the conventional membranes*) so they will require less power than used in processes leading to compression. CCS also investigates how to make an even more permeable nanotube, adhering at its ends, certain molecules to attract CO_2 but not other gases. This has been tried with other membranes, but those based on nanotubes have been most successful.

Metal-Organic Systems as New Materials for CO₂ Capture: These metal-organic systems (*metal-organic frameworks*) are lattices of organic compounds and metal atoms that have the advantage of offering a large internal surface area in which CO_2 molecules are captured. The lattices act as though they are crystal sponges. The metal used in these networks is magnesium, which produces the right environment to bind CO_2. An important advantage of this material is that 87% of CO_2 can be released at room temperature, releasing the remaining 13% by heating to about 75 °C, which is well below the temperatures required to regenerate to conventional chemical solvents.

Futuregen Project: The project is promoted by the US Department of Energy (Fig. **17**). It will be carried out in alliance with numerous companies from various countries [40].

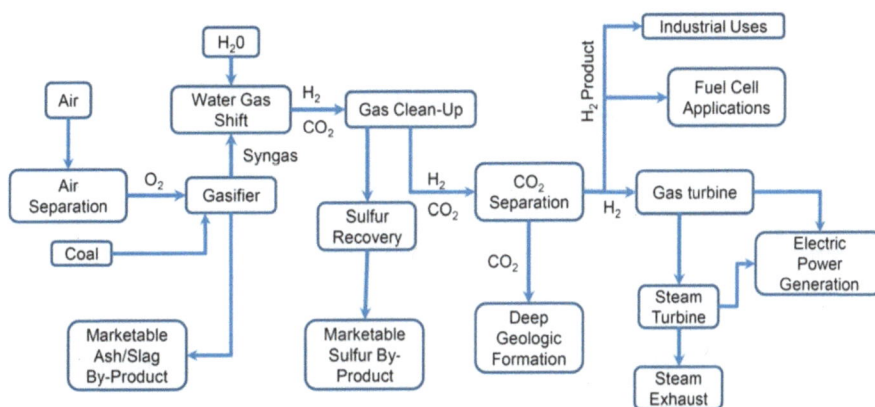

Figure 17: Schematic of IGCC plant integrated with CCS technology, for the FutureGen Alliance project.

The project aims is to build an electric power production plant at Mattoon, Illinois, with capacity of 275MW fueled by coal and provided with CCS for the virtual elimination of emissions which would be stored in a deep aquifer [41]. The plant will also produce large amounts of hydrogen. This has motivated the U.S. administration to temporarily suspend other hydrogen production related programs. The carbon gasification technique to be used is the so-called IGCC (Integrated Gasification power plants, in Combined Cycle) so that the CCS will be used for pre-combustion (Fig. **18**).

The future technology trends are established according to the great challenges that CCS has, if it is to play a meaningful role in the future in mitigating climate change. In CCS technologies, the biggest challenges lie in scaling the facilities by a factor of ten or more as well as in lowering energy expenditures, especially in CO$_2$ capture processes [42].

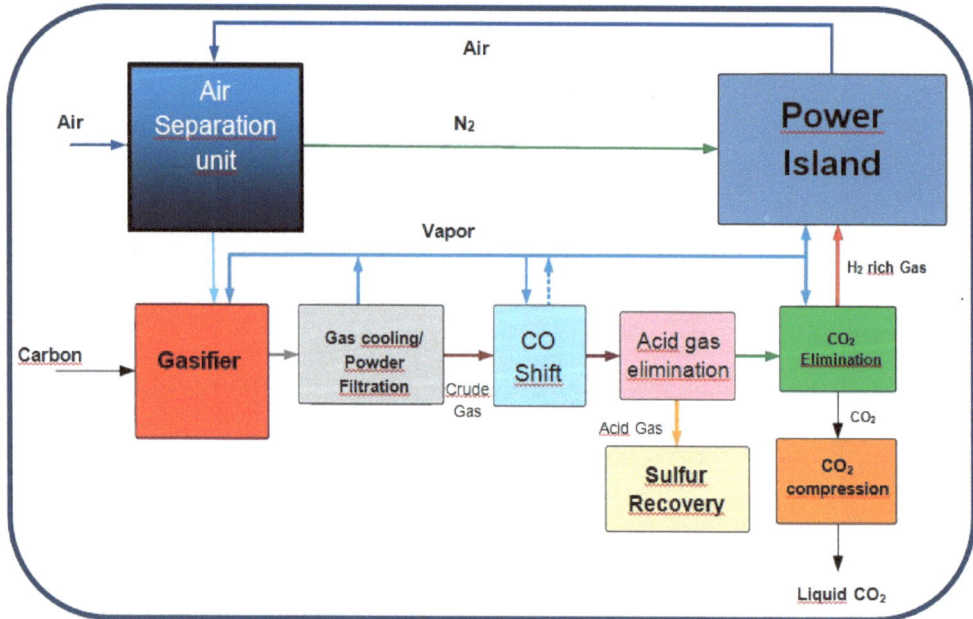

Figure 18: IGCC process with CO$_2$ capture and "shift" acidic conversion.

Furthermore, given that higher raising the efficiency of electric power plants, lowers the cost of capture and storage of CO$_2$ per kWh produced. Therefore, it is thought that new combined cycle plants using super alloys, high temperature hydrogen turbines, or more efficient techniques for separation of CO_2, could have the same cost of electricity as existing plants without CCS (Fig. **19**). Other important aspects that are being studied are the proper location for CCS in industrial facilities and the best way to integrate CCS processes.

The *FutureGen* project is also studying the storage of CO_2 in the ocean waters as in mineral carbonation, which still requires a great deal of evidence to ensure that there is no environmental harm, since it could increase the problem of ocean acidification. Given what it has done so far, it has been excluded as future technology at least by the European Union.

Finally, converting the gas by chemical reactions into a solid as incarbonationalso poses some important challenges. One is the large amount of reagent needed. Another is finding adequate places to store the large amount of reaction product generated.

8. FUTURE TECHNOLOGICAL DEVELOPMENTS BASED ON IBA

As can be seen from the information presented in previous sections, the current CCS-oriented technologies are developing rapidly. Development is conditioned on one hand by the need to *"alleviate"* the problem of climate change induced by environmental pollution gases that produce the *"greenhouse effect"* [43, 44]. CO_2 stands out among these pollutants because of the significant amount in which it is released into the atmosphere. Development is also conditioned by the search for greater efficiency in the use of primary energy sources while avoiding collateral environmental damages, among which could be damage of a social type.

One of the main remaining challenges is to determine a social-scientific approach to solve the aforementioned problems that allows the development of innovative technologies supported by basic research and driven by public policies.

In this context, IBA is considered to be a major contributor to open new approaches that may raise possibilities of technological developments in CCS. This is because the analytical capabilities of IBA have advantages such as:

    a)   *Creation of new materials by ion implantation;*

b) *Analysis capabilities with "micro and nano" spatial resolution, and parts per million in terms of concentrations, even reaching up to parts per trillion in pre-concentrated samples; and*

c) *The technique of "Accelerator Mass Spectrometry" (AMS) has a nominal accuracy of 1 atom, in 10^{15} atoms.*

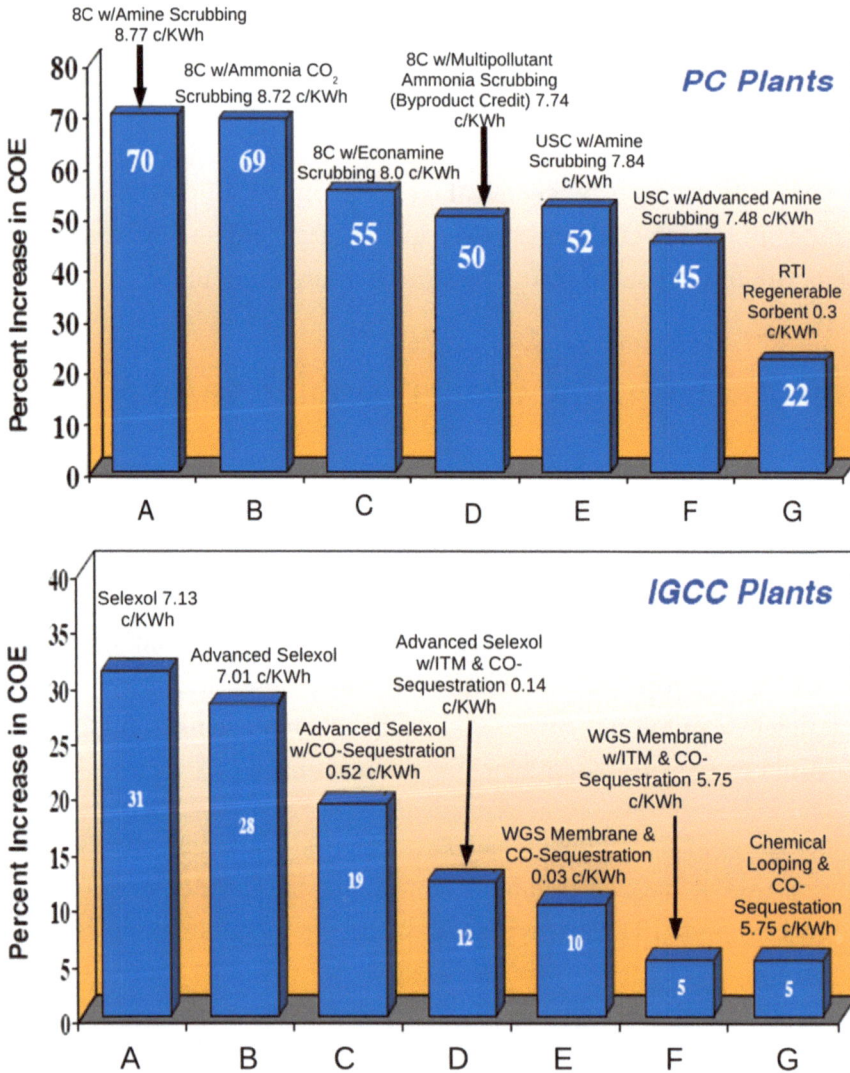

Figure 19: Cost of electricity (COE) increases for power plants with CO$_2$ capture and storage using current technology (column A) and various advanced technologies (columns B to G).

ACKNOWLEDGEMENTS

We thank very dutifully to Gina Ross and David Collins, their collaboration in preparing of this paper.

CONFLICT OF INTEREST

The authors confirm that this chapter contents have no conflict of interest.

REFERENCES

[1]   Intergovernmental Panel on Climatic Change (IPCC), Carbon Dioxide Capture and Storage (Special Inform), Technical Summary, [Online] **2005,** http://www.ipcc.ch/pdf/special-reports/srccs.html, [Accessed on: August 13th, 2013].

[2]   Tesmer J.R.; Nastasi, M. *Handbook of modern ion beam material analysis,* Materials Research Society, **1995.**

[3]   Fernández Niello J.; Testoni J. *Elementos de reacciones nucleares*, Instituto de Tecnología, J. Sábato, Buenos Aires, **1997.**

[4]   Centi, G; Perathoner, S. Opportunities and prospects in the chemical recycling of carbon dioxide to fuels. *Catal. Today.,* **2009,**148, 191-205.

[5]   Gale, J.; Davidson, J. *Transmission of CO₂. Safety and Economic Considerations.*6th Conference on Greenhouse Gas Control Technologies, Kyoto, pp 1-4, **2002.**

[6]   Royal Society of Chemistry (RSC), Converting CO₂ to chemicals, Workshop of the RSC: Environmental, sustainability and energy forum, Londres, **2006** (www.rsc.org/images/converting%20CO2%20to%20chemicals%20report_FINAL_tcm18-65202.pdf) [Accessed on: August 17th, 2013].

[7]   IEA GHG, 2000b. *Leading options for the capture of CO₂ emissions at power stations*;report PH3/14, IEA Greenhouse Gas R&D Programme, Cheltenham, UK. Feb 2000.

[8]   Elewaut E., Koelewijn D., van der Straaten R., Bailey H., Holloway S., Barbier J., Lindeberg E., Miller H., and Gaida K. *Inventory of the theoretical storage capacity of the European Union and Norway.* In The Underground Disposal of Carbon Dioxide (ed. S.Holloway). Final Report of the Joule II Project CT92-0031. British Geological Survey, **1996.**

[9]   Hendriks, C. *Carbon dioxide removal from coalfired power plants.* Dissertation, Utrecht University, Netherlands, **1994.**

[10]   Ishibashi, M.; Otake, K.; Kanamori S.; Yasutake, A. *Study on CO₂ Removal Technology from Flue Gas of Thermal Power Plant by Physical Adsorption Method,*//Ibid, pp. 95- 100.

[11]   Y. Takamura;Y. Mori; H. Noda; S. Narita; A. Saji; Y. S. Uchida; *Study on CO₂ Removal Technology from Flue Gas of Thermal Power Plant by Combined System with Pressure Swing Adsorption and Super Cold Separator*; Proceedings of the 5th International Conference on Greenhouse Gas Control Technologies, 2000 August 13-16; Williams D. eds.; CSIRO Publishing, Collingwood, Victoria, Australia, **2000.**

[12]   Centi, G. *Converting CO₂ back to fuel.* ACS 232nd National Meeting and Exposition, San Francisco, September **2006.**

[13]   Holloway, S.; Van der Straaten, R. The Joule II project. The Underground Disposal of Carbon Dioxide. *Energ Convers Manage*, **1995**, Vol. 36(6-9), 519-522.

[14]   Ohsumi T. Prediction of Solute Carbon Dioxide Behavior Around a Liquid Carbon Dioxide Pool on Deep Ocean Basins. *Energy Convers. Manage.*, **1997**, 34, 1059-1064.

[15]   Chapel, D.G.; Mariz, C.L.; Ernest, J. *Recovery of CO₂ from flue gases: commercial trends*; Annual Meeting of the Canadian Society of Chemical Engineering, Saskatoon, Canada, **1999**.

[16]   Mimura, T; Satsumi T. S.; Iijima, M.; Mitsuoka, S.; Development on Energy Saving Technology for Flue Gas Carbon Dioxide Recovery by the Chemical Absorption Method and Steam System in Power Plant, 4th International Conference on Greenhouse Gas Control Technologies.; P., Riemer, B., Eliasson, A., Wokaun, Eds.; Elsevier Science, Ltd., UK, **1999**,pp.71-76.

[17]   Mimura, T., T. Nojo, M. Iijima, T. Yoshinayama and H. Tanaka, 2003: Recent developments in flue gas CO₂ recovery technology. Greenhouse Gas Control Technologies, Proceedings of the 6th International Conference on Greenhouse Gas Control Technologias (GHGT-6), 1-4 Oct. 2002 Kyoto, Japan, J. Gale And Y Kaya (eds). Elsevier Science Ltd, Oxford, UK.

[18]   Yokoyama, T. *Japanese R&D on CO₂ Capture*. Proceedings of the 6th International Conference on Greenhouse Gas Control Technologies (GHGT-6), 2002, October 1-4; J. Gale, Y. Kaya, Eds.; Elsevier Science Ltd, Oxford, UK., **2002**, pp. 13-18.

[19]   Chaffee, A. L.; Knowles, G. P.; Liang, Z.; Zhang, J.; Xiao, P. y Webley, P. A. CO₂ capture by adsorption: materials and process development. *International Journal of Greenhouse Gas Control* 1, **2007**, 11.

[20]   Hori, H.; Takano, Y.; Koike, K.; Sasaki, Y. Efficient rhenium-catalyzed photochemical carbon dioxide reduction under high pressure. *Inorg. Chem. Commun.* 6, **2003**, 300.

[21]   Kaneko, S.; Shimizu, Y.; Ohta, K.; Mizuno, T. Photocatalytic reduction of high pressure carbon dioxide using TiO₂ powders with a positive hole scavenger. *J. Photoch. Photobio* A,115, **1998**, 223.

[22]   Shi, D.; Feng, Y.; Zhong, S. Photocatalytic conversion of CH4 and CO₂ to oxygenated compounds over Cu/CdS-TiO₂/SiO₂ catalyst. *Catal. Today* 98, **2004**, 205.

[23]   Mathieu, P. Mitigation of CO₂ emissions using low and near zero CO₂ emission power plant, *International Journal on Energy for a Clean Environment*, **2003**,4, 1-16.

[24]   The Zero Emission Fossil Fuel Power Plants Technology Platform, European Commission, Community research. "A Vision for Zero Emission Fossil Fuel Power Plants". EUR 22043, 2006.pp 19-28.

[25]   Abu-Zahra, M.R.M.; Schneiders, L.H.J.; Niederer, J.P.M.; Feron, P.H.M.; Versteeg, G.F. CO₂ capture from power plants. Part I. A parametric study of the technical performance based on monoethanolamine. *Internattional Journal of Greenhouse Gas Control* 1, **2007**, 37.

[26]   DOE. *Carbon sequestration Research and Development*. Office of Fossil Energy, NETL, US. Department of Energy, **1999**.

[27]   Doctor, R.; Molburg, J.C.; Brockmeier, N.F. *Transporting Carbon Dioxide Recovered from Fossil-Energy Cycles*. 5th International Conference on Greenhouse Gas Control Technologies, Cairns, **2000**, pp. 567-571.

[28]   Wilkinson, M.B.; Simmonds, M.; Allam, R.J.; White, V.; *Oxy-fuel conversion of heaters and boilers for CO₂ capture,* 2nd Annual conference on Carbon Sequestration, **2003**.

[29]   Dillon, D.J.; Panesar, R.A.; Wall, R.A.; Allam, R.J.; White, V.; Gibbins, J.; Haines, M.R.; *Oxycombustion processes for CO₂ capture from advanced supercritical PF and NGCC power*

plant, Proceeedings of 7th international conference on greenhouse gas control technologies, **2005**, 1.

[30]   Odenberger, M.; Svensson, R. *Transportation Systems for CO₂ – Application to Carbon Sequestration.* Technical report T2003-273. Department of Energy Conversion Chalmers University of Technology, Suiza, 2003.

[31]   Office of Scientific and Technical Information (OSTI), Carbon sequestration research and development: advanced chemical approaches to sequestration, **2007** (www.osti.gov/energycitations/servlets/purl/8107229s7bTP/native/810722.pdf) [Accessed on: August 20st, 2013].

[32]   Reiche, D. E.; Houghton, J. C.; Kane, R.L. DOE, A review of carbon sequestration science and technology opportunities by the US Department of Energy, **2007** (www.lib.kier.re.kr/balpyo/ghgt5/Papers/E1%202.pdf) [Accessed on: August 19th, 2013].

[33]   Naqvi, R.; Bolland, O., Multi-stage chemical looping combustión (CLC) for combined cycles with CO₂ capture, *International Journal of Greenhouse Gas Control*, **2007**, 1.

[34]   F. de Diego, L; García-Labiano, F.; Gayán, P.; Celaya, J.; Palacios, J.M.; Adanes, J.; Operation of a 10 kWth chemical-looping combustor during 200 h with a CUO-Al2O3 oxygen carrier, *Fuel*, 86, **2007**.

[35]   Anpo, M.; Yamashita, H.; Ichihashi, Y.; Fujii, Y. y Honda, M. Photocatalytic reduction of CO₂ with H2O on titanium oxides anchored within micropores of zeolites: effects of the structure of the active sites and the addition of Pt. *Phys. Chem.* B, **1997**, 101, 2632-2636.

[36]   Gou, C.; Ruixian C.; Zhang G.; An advanced zero emission power cycle with integrated low temperature thermal energy, *ApplTherm Eng.*, **2006**, 26.

[37]   Linsky, J.; Karjalainen, T.; Pussinen, A.; Narbuurs, G.J.; Kaupi, P. Trees as Carbon Sinks and Sources in the European Union. *Enviromental Science and Policy* 3, **2000**, pp. 91-97.

[38]   Houghton, J.; Ding, Y.; Griggs, D.; Noguer, M.; Van der Linden, P.; Dai, X.; Maskell, K.; Johnson, C. *Climate Change 2001: The Scientific Basis.* Cambridge University Press. IEA GHG. **2001**.

[39]   Usubharatana, P.; McMartin, D.; Veawab, A.; Tontiwachwuthikul, P. Photocatalytic process for CO₂ emission reduction from industrial flue gas stream. *Ind. Eng. Chem. Res.* 45, **2006**, 2558.

[40]   Edwards, J.H. Potential sources of CO₂ and the options for its large-scale utilisation now and in the future. *Catal. Today*, **1995**, 23, 59-66.

[41]   Marin, O.; Bourhis, Y.; Perrin, N.; DiZanno, P.; Viteri, F.; Anderson, R.; *High efficiency zero emission power generation based on a high temperature steam cycle*, 28th. International Technical Conference on Coal Utilization and Fuel Systems, **2003**.

[42]   Riemer, P.W.F.; Ormerod, W.G. International perspectives and the results of carbon dioxide capture disposal and utilisation studies; *Energ Convers Manage,* 36(6-9), 813-818. Riemer y Ormerod, **1995**.

[43]   Jose D. Figueroa, Timothy Fout, Sean Plasynski, Howard McIlvried, Rameshwar D. Srivastava. "Advances in CO2 capture technology—The U.S. Department of Energy's Carbon Sequestration Program". Elsevier (eds), *Int. J. Greenh. Gas Control* 2., 2008, 9–20.

[44]   IPCC. *Climate Change 1995: The Science of Climate Change.* Ed. Cambridge University Press, Cambridge, UK.H., **1996**.

Index

A

ABB Lummus Crest MEA process 70

Absorber 70, 71, 73, 74, 85, 86, 103, 104, 106, 112, 228, 232-36

Absorption 40-42, 50-52, 56, 64, 70-76, 78, 85, 87, 89, 90, 95-97, 102-6, 110-12, 147, 148, 159, 160, 162, 163, 167, 178-83, 185, 187, 191, 193, 196, 203, 205, 206, 210-13, 215, 219, 220, 224-29, 233, 236, 245, 254, 259
   chemical 41, 179, 206
   maximum CO_2 95, 97

Absorption column 72, 104, 110, 111, 147, 159, 167, 236

Absorption efficiency 111, 225, 227, 228

Absorption of CO_2 76, 78, 106, 162, 167, 178-81, 183, 187, 205, 219, 220, 224-28, 236, 259

Absorption processes 95, 160, 162, 180, 182, 191, 193, 254
   chemical 40
   downstream CO_2 87
   large scale 96

Absorption segments 178, 210-12, 215

Absorption systems 85, 89

Activity coefficients 219-22
   asymmetric 221, 222

Adiabatic 211

Adsorbent 256, 257

Adsorption 71, 72, 96

Air-firing 3, 6, 19, 20, 24-27, 29, 32

Alloys 156, 157

American Society of Testing Materials (ASTM) 147, 159

Amines 41, 69, 72, 74, 96, 104, 106, 122, 179, 181, 183, 256
   primary 72, 97
   ternary 72

Ammonia 178, 179, 181, 183, 218, 220, 225, 228, 229, 232, 233, 236
   molar fraction of 225, 227, 228
   mole fraction of 226

C

www.ingramcontent.com/pod-product-compliance
Lightning Source LLC
Chambersburg PA
CBHW050811220326
41598CB00006B/182